Suzuki, Ueta, Mori • Global Environmental Security

Springer

Berlin
Heidelberg
New York
Barcelona
Budapest
Hong Kong
London
Milan
Paris
Santa Clara
Singapore
Tokyo

Y. Suzuki • K. Ueta • S. Mori (Editors)

Global Environmental Security

From Protection to Prevention

With 59 Figures

Springer

YUJI SUZUKI
Hosei University, Japan
17-1, 2-chome Fujimi, Chiyoda-ku,
Tokyo 102 Japan

KAZUHIRO UETA
Kyoto University, Japan
Yoshidahon-cho, Sakyo-ku,
Kyoto-shi, Kyoto 606 Japan

SHUNSUKE MORI
Science University of Tokyo, Japan
Yamasaki 2641, Noda-shi,
Chiba 278, Japan

ISBN-13:978-3-642-80154-9

Cataloging-in-Publication Data applied for
Die Deutsche Bibliothek - CIP-Einheitsaufnahme
Global environmental security : from protection to prevention / J. Suzuki ; K. Ueta ; S. Mori. - Berlin ; Heidelberg ; New York ; Barcelona ; Budapest ; Hong Kong ; London ; Milan ; Santa Clara; Singapore ; Tokyo : Springer 1996
ISBN-13:978-3-642-80154-9 e-ISBN-13:978-3-642-80152-5
DOI: 10.1007/978-3-642-80152-5
NE: Suzuki, Juji [Hrsg.]

Softcover reprint of the hardcover 1st edition 1996

Typesetting: camera-ready by authors

SPIN: 10531956 32/3136 - 5 4 3 2 1 0 - Printed on acid-free paper

Foreword

Z. Sonoda

It is my great honour to declare the opening of the 2nd IBM Japan International Symposium on 'The Global Environment'.

It is especially my honour, on behalf of IBM Japan, to welcome all of you most distinguished experts of the world.

Historically, IBM Japan has maintained a strong interest in the protection of the global environment, not only as a member of the industrial community, but also as a member of the global community where environmental disruption has become more and more significant and of common concern.

In 1990, we held the first environmental symposium on 'Global Warming'. It offered a stage to which both social and natural scientists could bring problems about 'The Global Environment' for in-depth discussions. One of its conclusions includes collaboration about socio-economic aspects among the industrialized and developing countries. With this in mind, we now host the 2nd symposium which focuses upon a socio-economic approach to global environmental problems.

I understand that the symposium will seek a solution by examining actual experiences and circumstances from various parts of the world, with special attention given to North/South issues. Thus, we have invited guests from America, Europe, and neighbouring Asian countries alike.

I wish, and believe, that the symposium will bring about new clues, toward a breakthrough for the betterment of the environment through positive and harmonious discussions in this beautiful setting at the foot of Mt Fuji.

For more than 20 years, IBM Japan has been conducting contributory projects in various areas of need at all levels of the human community—local, regional, national and global. The social needs areas we address are welfare for the disabled, medical problems of incurable diseases, promotion of basic science, enhancement of educational and cultural infrastructure, and global environmental problems. This symposium is held within this framework, on the company's long-term commitment

I would therefore like to again express my deepest appreciation to all the honourable participants, in particular to the executive committee members, Professor Suzuki of Hosei University, Professor Mori of the Science University of Tokyo, and Professor Ueta of Kyoto University who all played a massive and integral role in orchestrating this event with the advisers—Dr Kondo of the Science Council of Japan and Professor Takeuchi of the University of Tokyo—who oversaw the framework and gave very important suggestions and effective support for the symposium.

Although Dr Kondo is in Russia at this time, I have the pleasure of introducing Professor Kawata, Vice-President of the Science Council of Japan.

Preface

Since the 2nd IBM Japan International Symposium on the Global Environment, September 20–22, 1993 in Gotenba, Japan, a number of new developments have taken place on the issue of global environment. First of all, the enthusiasm that made the Earth Summit of Rio de Janeiro in 1992 a historic success has greatly subsided. Instead, global concerns have drifted from the protection of global environment to trade negotiations and ethnic conflicts. Mesmerized by such rapid evolvements post Cold War, the world seems to have almost no positive development toward environmental protection in sight.

Secondly, however, we have witnessed positive developments as well. Many more local authorities at sub-national level have taken concrete steps toward environmental protection. Some municipal governments, including those in the Third World, have put into practice a set of environmental policies such as Agenda 21, adopted at Rio de Janeiro. Quite a few of them, in addition, have enhanced law enforcement measures so that environmental policies can be more systematically administered. Such initiatives at this level are important since they enable a greater participation of local population in the policy making and policy implementation processes.

Thirdly, the transnational cooperation for environmental protection has been greatly reinforced, especially by the extensive and intensive activities of the so called 'green' non-government organizations. Today, all governments across nations are under increasing demand for environmental protection. In addition, those concerned people have strengthened their mutual cooperation at grassroot level. And such transnational cooperation has contributed to promote or even force governments—local, national or international—to take further steps toward the protection of the global environment.

These developments suggest the discussion of environmental protection is undergoing a serious change. The era for agenda setting has ended, and a new era has arrived for concrete action. This change is seen clearly in the efforts of the IBM Japan International Symposium on the Global Environment, too. Its first symposium of 1991, whose main participants were natural scientists from the North, discussed global warming extensively. The second meeting, in 1993, saw more officials and social scientists from North and South and sought ways and means for the protection of the global environment. As you will see, the discussions following often cut across conventional divisions between natural and social sciences, theories and practices, governments and non-government organizations, and even North and South.

This is the record of the 2nd IBM Japan International Symposium. We hope this volume contributes to the enhancement of common understanding and action for the protection of our globe. Last but not least, all the participants are most appreciative of IBM Japan. Without that company's support, this symposium would not have taken place. Responsibilities for the analyses as well as arguments included in this volume belong with the authors and the editors.

The Editors, March 1995

IBM Japan
International Symposium on the Global Environment

Advisory Committee:	J. Kondo	Science Council of Japan
	K. Takeuchi	University of Tokyo
Executive Committee:	Y. Suzuki	Hosei University
	K. Ueta	Kyoto University
	S. Mori	Science University of Tokyo
Secretariat:	Z. Sonoda	IBM Japan
	A. Nemoto	IBM Japan

Opening Remarks

Tadashi Kawata

Vice President (Cultural & Social Sciences) Science Council of Japan

Mr Chairman, distinguished guests, ladies and gentlemen:

On behalf of the Science Council of Japan, it is a great pleasure for me to have this opportunity to extend warm greetings to all of you who have gathered here from various countries at this opening session of the 2nd IBM Japan International Symposium on 'The Global Environment'.

As is well retained in our memory, the United Nations Conference on the Human Environment was held in Stockholm in 1972. To commemorate the 20th anniversary of the Stockholm Conference, the United Nations Conference on Environment and Development, known as the Earth Summit, was held in Rio de Janeiro in 1992. That event brought together more than 150 nations, 1,400 non-governmental organizations and 8,000 journalists.

The efforts and experiences in the promotion of international cooperation and exchange on environmental problems over these past 20 years have been invaluable. The Earth Summit should be judged in the context of a process of increasing attention, practical measures and effectiveness in the management of environmental issues. Obviously, the Earth Summit was not an end but a new beginning, as the Summit's Secretary-General, Maurice F. Strong, said in his opening address to that conference.

Thus, an agenda for follow-up actions at the national, regional and international levels will demand commitments on the part of all governments and non-governmental organizations, and academic involvement on the part of both social and natural scientists.

It is now acknowledged by all, that the environment is not the business of the natural scientists alone. Human actions have been responsible for the deteriorating state of the global environment; similarly, changes in the global environment are now influencing human activity. The earth and its human inhabitants are interacting partners in a single system. The future of the environment, and of human societies, will be jeopardized if adequate action is not taken and the path of sustainable development is not chosen.

I believe that it is vitally important for social scientists to look long and hard at global change, and hold further dialogue with natural scientists. Perhaps there may be differences in approach between them, but, in dealing with global environmental problems, it is essential that they talk with each other and work together.

From this perspective, I think it is really fortunate that the program of this IBM Symposium has been so well formulated, with attention to detail, and with the full understanding that environmental problems are the products of technological, economic, socio-political and administrative development.

In addition, it seems to me that the emphasis of this Symposium has been placed upon reviewing and applying the findings of research and the current state of knowledge, with a particular emphasis on policy relevance. I believe that this kind of approach should be considered essential in strengthening and redoubling our cooperative efforts towards environmental protection across the world.

The proposed topics of this Symposium—say, 'Technology for Sustainable Development', 'Toward Establishing a Global Environmental System', 'New Initiatives for Environmental Protection', and so forth—seem to be very attractive subjects. Hence, I am now contemplating staying here, and participating in this Symposium as an observer from tomorrow morning—first session. Perhaps Professor Yuji Suzuki, an important member of the executive committee of this Symposium may permit me to do so, as he is a very close young colleague of mine.

The Science Council of Japan was established in 1949, soon after the Second World War, comprising 210 members with three year terms of office, representing all fields of science. The Council acts as the representative of Japanese scientists, both domestically and internationally, for the promotion of science. The Council is attached to the Prime Minister's office, but it functions independently.

As for international scientific exchanges, the Council conducts such activities as affiliating with international scientific organizations, sending delegations to international congresses held overseas, and organizing and supporting important international conferences in Japan.

At the end of 1992 the Council was a member of 46 international umbrella organizations, including the International Council of Scientific Unions (ICSU) and the International Federation of Social Science Organization (IFSSO). Since 1983, the Council has been executing the Bilateral Scientific Exchange Programs, which involve sending delegations abroad in order to promote mutual understanding in scientific fields with other countries, through discussions with leading scientists and policy makers.

Incidentally, we are now sending our delegation headed by Dr Jiro Kondo, President of the Council, to Russia. He could not come here today for that reason and has sent his apologies.

In recent years there has been an increase in international research programs on subjects requiring consideration on a global scale, such as research on solar terrestrial energy, atmospheric phenomena, environmental problems and so on.

As for environmental problems, our Council adopted the recommendations regarding the International Geosphere-Biosphere Program (IGBP) during its general meeting held in April 1990 and urged the Japanese government to offer positive support to the IGBP study activities. At the same time, the Council has set up the Japan National Committee for IGBP (IGBP-Japan) and organized the IGBP study plan which has focused on the regions of Monsoon Asia and the West Pacific. Subsequently, the Council set up the Japan National Committee of the Human Dimensions of Global Environmental Change Program (HDP) in 1991.

There is an umbrella organization called Association of Asian Social Science Research Councils (AASSREC) in the Asia-Pacific region. Recently our Council affiliated with AASSREC and complied with the request by the AASSREC Secretariat to host the tenth Biennial Conference of AASSREC in Japan in 1993. Early in this month this conference was held in the city of Kawasaki, and we had a Symposium on 'Environment and Sustainable Development: Social Science Perspectives', which focused on how the environmental situation is viewed from the vantage point of the Asia-Pacific region.

The Science Council of Japan conducts deliberations on important matters concerning science, and makes efforts to implement any decision reached. The Japanese government may seek the opinions of the Council and the Council has the right to take the initiative in making recommendations to the government.

At the request of the government last April, the Council adopted the recommendations concerning contributions to the international community in scientific fields at its general meeting. These recommendations were submitted to the government (Prime Minister's office). The following items may perhaps be regarded as the important specific recommendations:

I. Upgrading the channels for the outflow of information from Japan as well as strengthening the interlinking of scientific research in the Asia-Pacific region.
II. Upgrading the support of international scientific exchanges, including the support of visiting researchers and foreign students.
III. Building a new system for international contributions in scientific fields.

I have brought here the full texts of these recommendations for foreign guests. Our Council intends to pursue its deliberations on important matters in the future; and, in connection with that, it will solicit the opinions and suggestions of scientists and people involved in related organizations in Japan and overseas.

The presentations of various views and observations on global environmental change at this Symposium will not only enrich the overall knowledge base, but will also help greatly in establishing priorities for future work. I sincerely hope all of the participants have a lively and productive discussion based on the presentations, make great achievements and create a profitable result from the 2nd IBM Japan International Symposium. Thank you very much.

Introduction

Yuji Suzuki
Hosei University, Tokyo, Japan

In retrospect, the UN Conference on the Human Environment in Stockholm, June 1972, earmarked a turning point in the history of economic development and the environment. It is not merely because it was the first international conference, in which 113 nations participated, that was specifically dedicated to discussions on the global environment. It is also because the conference set a basic tone to policy making and policy implementation for environmental protection. The Declaration on the Human Environment and the Action Program, both adopted at the conference, have remained as the standard references until today.

Since Stockholm, a number of environmental conferences have been organized and nearly 60 treaties, conventions and agreements signed. Yet the decline of the global environment continues, the problems increase, the damages accumulate—and, in addition, the needed collaboration across the world seems more difficult than ever before. Against this backdrop, nearly 180 nations took part in the UN Conference on Environment and Development in Rio de Janeiro, June 1992, and the concrete action program called Agenda 21 was signed. The environment has become a more acute and common concern among nations, especially in the post Cold War period.

Nonetheless, not much has been changed and the desired collaboration has got bogged down. Following the 'tri-crisis' of the global environment, namely population explosion, the accelerated consumption of natural resources, and massive poverty, we now face the North/South division, too. Particularly sharp divisions have been drawn with respect to (1) the responsibility of the North in relation to the degradation of global environment, (2) the right of development in the South, and (3) the financing of sustainable development as a compensation for the environmental damage or as an addendum to the existing ODA mechanism. Despite mutual concern for the global environment, the North/South division has become a stonewall to the much needed global cooperation.

Yet, the 'tri-crisis' does not excuse a lack of action for the protection of the global environment. Environmental degradation is a fact of daily life at the grassroots and it is at this level that concrete actions are urgently needed. What we need now is to set into practice those international or national instruments which we have obtained in the past 20 years or so. In addition, action at the grassroots, if desired to be effective, has to mobilize all available ways and means—including international aid. These actions should be buttressed by a long-term policy for the promotion of sustainable development. To cope with this demand, we really need a holistic way of thinking on the environment, comprehensive policy making and implementation, and multi-dimensional cooperation across the board.

With this change in perspective, the symposium deliberately started with the conceptual transformation of the global environment by inviting speakers both from the North and the South. The papers and discussions are included in Chapter One. Then the three issue areas were selected for scrutiny, based on past experience as well as for the identification of obstacles. The three issues are (1) the possibility of technological breakthrough, (2) the economy of environmental protection, and (3) the search for international cooperation at the grassroots. The exploration of these issues is undertaken in Chapters Two, Three, and Four, respectively. After detailed examination of the selected areas, the symposium discussed new initiatives for the protection of the global environment. This is Chapter Five. And finally, at the closing session the symposium discussed 'A Scenario for the Future' which was adopted and is included in the Conclusion.

Contents XV

Chapter One

The Conceptual Transformation of the Environment toward Global Cooperation for Environmental Protection

I.1 Overview

Chapter One contains three separate papers and one appraisal remark. Different conclusions and recommendations are drawn by each presenter, based on their own interests, expertises and conceptions of environment. However, what is important here is the fact that all of them stress an urgent need to transform the concepts of environment, development and technology.

The first paper, by Lester R. Brown, argues that the time has come to launch the 'Environmental Revolution'. As opposed to the two earlier great changes, viz the Agricultural Revolution and the Industrial Revolution, its success depends not so much on technological advances as on restructuring the global economy. He examines the pivotal role that governments should play through tax policies, subsidies, regulations, R&D funding, and procurement policies in order to squeeze centuries of change into decades.

The second paper, by Emil Salim, begins with a critical review of the conventional development model. From the viewpoint of the South, he argues, this has led many Third World nations to the dilemma of development and environment. He then makes a search for a sustainable development model which can meet economic, social, political and ecological needs in the South, but also promote global cooperation between North and South.

The third paper, by Takao Kashiwagi, discusses the possibility of technological breakthrough and global cooperation. Examining the current environment-friendly clean energy and technologies, he stresses the need for a holistic approach to energy conservation including cascaded energy utilization, co-generation, and unexploited energy utilization systems. He then proposes global cooperation toward conceptual transformation for rationalized clean energy utilization and conservation.

In his appraisal, Kei Takeuchi discusses a transnational system which could guarantee cooperation across nations. The end of World War II unleashed dramatic changes in world politics. Yet, he observes, policy orientations among nations have shifted away from environmental protection to other issues. Against this backdrop, he argues, firmer steps should be taken—such as environment-keeping operations by the United Nations, coordination of aid programs and technology transfers.

In order to promote more concrete steps, however, one needs to tackle the following two problems. The first is the difficulty in conceptualizing what are the environmental problems. There are cases that appear to be problematic at one level but not at other levels. A typical example is the case of CFC. Due mainly to the fact that CFC is chemically stable, it was thought to be good. It did not appear to harm local environment. However, it seriously damaged the ozone layer. Another example is global warming. It may seem to be beneficial to some parts of the world in the short term, but in the long run it will certainly damage the life of the whole globe.

Regarding this problem, I believe there is a gross misunderstanding here. Even though the reality of international politics more often than not tends to prevent the international cooperation absolutely necessary for the protection of the global environment, no one can deny the fact that the world has become increasingly small.

The second problem I should point out here is the contradiction between economic growth and environmental protection. Most often these are seen as contradictory and presented as a matter of choice.

I.2 Launching the Environmental Revolution

L. R. Brown

Worldwatch Institute, 1776 Massachusetts Ave, NW. Washington, DC 20036–1904, USA

Abstract

The 1990s will be seen as a hinge point in history. The degradation of the environment which is now taking place cannot continue without threatening civilization itself. Building an environmentally sustainable future depends on restructuring the global economy so it does not destroy its natural support systems, including major shifts in human reproductive behavior, values and life-styles. This will result in a revolution on the scale of the agricultural and industrial revolutions.

Governments must play a big role in the environmental revolution. The partial replacement of income and other taxes with environmental taxes could be their most effective tool. Their challenge will be to deal simultaneously with poverty and environmental degradation, so cooperation between countries is essential. The other major role-players will be large corporations, which have been resistant to the environment revolution. Short-term profits have taken precedence over long-term environmental concerns. Corporations can, however, be encouraged to change through taxes and by an environmentally concerned market.

If the environmental revolution is to succeed, it will need support and active participation from all.

1 The Environmental Revolution

When the history of the late twentieth century is written, the 1990s will be seen as a decade of discontinuity—a time when familiar trends that had seemed likely to go on forever, like smooth straight roads climbing toward an ever receding horizon, came to abrupt bends or junctures and began descending.

I am happy to be here speaking to you at a time that historians will call a hinge point of history, and I am grateful for the invitation. Not only are many trends changing today, but many people are making them change. Those of us here in this room will be among those people.

At Worldwatch, as we work on the eleventh *State of the World* book, we are reporting essentially the same trends that we have been reporting in the 10 preceding issues. Each year, the earth's tree cover shrinks. Each year, the deserts grow larger. Each year, we report that the topsoil on a third of our crop land is getting thinner from erosion. Each year, we report that the concentration of greenhouse gases in the atmosphere is building in an all too predictable fashion. Each year, we report that the stratospheric ozone layer is being depleted, and now at an accelerating rate. Each year, we report that the number of plant and animal species with which we share the planet is diminishing.

I think that most of us would agree that these trends cannot continue indefinitely without threatening civilization itself. Our world faces potentially convulsive change. The question is, in what direction will it take us? Will the change come from strong worldwide initiatives that reverse the degradation of the planet and restore hope for the future, or will it come from continuing environmental deterioration that leads to economic decline and social instability?

Muddling through will not work. Either we turn things around quickly or the self-reinforcing internal dynamic of the 'deterioration and decline' scenario will take over. The policy decisions we make in the years immediately ahead will determine whether our children live in a world of development or decline.

There is no precedent for the change in prospect. Building an environmentally sustainable future depends on restructuring the global economy, major shifts in human reproductive behavior, and dramatic changes in values and life-styles. Doing all this quickly adds up to a revolution, one defined by the need to restore and preserve the earth's environmental systems. If this 'environmental revolution' succeeds, it will rank with the agricultural and industrial revolutions as one of the great economic and social transformations in human history.

Like the agricultural revolution, it will dramatically alter population trends. While the former set the stage for enormous increases in human numbers, this revolution will succeed only if it stabilizes population size, re-establishing a balance between people and the natural systems on which they depend. In contrast to the industrial revolution, which was based on a shift to fossil fuels, this new transformation will be based on a shift away from fossil fuels.

The two earlier revolutions were driven by technological advances—the first by the discovery of farming and the second by the invention of the steam engine, which converted the energy in coal into mechanical power. The environmental revolution, while it will obviously use new technologies, will be driven primarily by the restructuring of the global economy so that it does not destroy its natural support systems.

The pace of the environmental revolution will be far faster than that of its predecessors. The agricultural revolution began some 10,000 years ago and the industrial revolution has been underway for two centuries. But if the environmental revolution is to succeed, it must be compressed in to a few decades.

Progress in the agricultural revolution was measured almost exclusively by the growth in food output that eventually enabled farmers to produce a surplus that could feed city dwellers. Similarly, industrial progress was gauged by success in expanding the output of raw materials and manufactured goods. The environmental revolution will be judged by whether it can shift the world economy on to an environmentally sustainable development path, one that leads to greater economic security, healthier life-styles and a worldwide improvement in the human condition.

Many still do not see the need for such an economic and social transformation. They see the earth's deteriorating physical condition as a peripheral matter that can be dealt with by minor policy adjustments. But 20 years of effort have failed to stem the tide of environmental degradation. There is now too much evidence on too many fronts to take these issues lightly.

Already the planet's degradation is damaging human health, slowing the growth in world food production, and reversing economic progress in dozens of countries. Data released by the Russian Academy of Medical Sciences show 11% of Russian infants suffering from birth defects. With half the drinking water and a tenth of the food supply contaminated, 55% of school age children suffer health problems. Economic losses associated with the deterioration of Europe's forests from acid rain total $30.4 billion each year, roughly equal to the annual output of the German steel industry. These examples, and countless others, show that our health is closely linked to that of the planet.

A scarcity of new crop land and fresh water plus the negative effects of soil erosion, air pollution and hotter summers on crop yields are slowing growth of the world grain harvest. Combined with continuing rapid population growth, this has reversed the steady rise in grain output per person that the world had become accustomed to. Between 1950 and 1984, the historical peak year, world grain production per person climbed by nearly 40%. Since then, it has fallen roughly 1% a year, with the drop concentrated in poor countries. With food imports in these nations restricted by rising external debt, there are far more hungry people today than ever before.

On the economic front, the signs are equally ominous: soil erosion, deforestation and overgrazing are adversely affecting productivity in the farming, forestry and livestock sectors, slowing overall economic growth in agriculturally based economies. A United Nations (UN) assessment of the earth's dry land regions estimated that the degradation of irrigated crop land, rain fed crop land, and range land now costs the world more than $42 billion a year in lost crop and livestock output. The World Bank reports that after three decades of broad based economic gains, incomes fell during the 1980s in more than 40 developing countries. Collectively, these nations contain more than 800 million people—almost three times the population of North America and nearly one sixth that of the world. In Nigeria, the most populous country in the ill-fated group, the incomes of its 123 million people fell a painful 29%, exceeding the fall in US incomes during the depression decade of the 1930s.

Anyone who thinks these environmental, agricultural and economic trends can easily be reversed need only look at population projections. Those of us born before the middle of this century have seen world population double to five billion. We have witnessed the environmental effects of adding 2.5 billion people, especially in the Third World. We can see the loss of tree cover, the devastation of grasslands, the soil erosion, the crowding and poverty, the land hunger and the water pollution associated with this addition. But what if 4.7 billion more people are added by 2050, over 90% of them in the Third World, as now projected by UN demographers?

The decline in living conditions that was once predicted by some ecologists from the combination of continuing rapid population growth, spreading environmental degradation and rising external debt has become a reality for one sixth of humanity. Moreover, if a more comprehensive system of national economic accounting were used—one that incorporated losses of natural capital, such as topsoil and forests, the destruction of productive grasslands, the extinction of plant and animal species, and the health costs of air and water pollution, nuclear radiation and increased ultraviolet radiation—it might well show that most of humanity suffered a decline in living conditions during the 1980s.

Today we study the archaeological sites of civilizations that were undermined by environmental deterioration. The wheat lands that made North Africa the granary of the Roman Empire are now largely desert. The early civilizations of the Tigris-Euphrates Basin declined as the waterlogging and salting of irrigation systems slowly shrank their food supply. And the collapse of the Mayan civilization that flourished in the Guatemalan lowlands from the third century BC to the ninth century AD may have been triggered by deforestation and soil erosion.

No one knows for certain why the centers of Mayan culture and art fell into neglect, nor whether the population of one to three million moved or died off, but recent progress in deciphering hieroglyphs in the area adds credence to an environmental decline hypothesis. One of those involved with the project, Linda Schele of the University of Texas, observes: 'They were worried about war at the end. Ecological disasters, too. Deforestation. Starvation. I think the population rose to the limits the technology could bear. They were so close to the edge, if anything went wrong, it was all over.'

Whether the Mayan economy had become environmentally unsustainable before it actually began to decline, we do not know. We do know that ours is.

2 Mechanisms of Change

Converting to an environmentally sustainable economy in the time available depends on accelerating the process of change. But what are the mechanisms of change?

Social change occurs in response either to new information or to new experience. Atmospheric chemists, for instance, reported in the mid-1980s that the chlorofluorocarbons (CFCs) used in air conditioning, refrigerators and plastic foams were evaporating into the upper atmosphere and damaging the stratospheric ozone layer that protects us from dangerous ultraviolet radiation. As a result, the use of CFCs is now being phased out. None of us has ever seen the ozone layer, much less observed its depletion. Nonetheless, the world responded—a classic case of new information leading to policy changes.

Six years ago, CFC manufacturers were opposed to the suggested phase out, regularly challenging the scientific evidence, much as coal producers are now questioning the science underlying the greenhouse effect. But when a few leading producers considered what would be in the long-term interest of their companies, they moved rapidly to gain the lead in developing alternatives. In a similar fashion, once a leading oil or coal company starts to invest heavily in the solar-hydrogen economy, the heir apparent to fossil fuels, others are likely to follow.

Sometimes new information alone will not bring about change. A smoker bombarded with data every few weeks from yet another study on the adverse health effects of this addiction may continue to smoke cigarettes, rationalizing away their life-threatening quality. But the day that person experiences difficulty breathing and is diagnosed with lung cancer, they will almost certainly stop smoking. In this instance, new experience did what new information could not.

To what extent will new information alone sustain the environmental revolution and to what extent, as with the smoker, will it take some painful new experience? With nuclear power, some governments elected not to invest in this energy source based on the studies that warned of the risk of potentially catastrophic accidents. Others rejected nuclear power only after the May 1986 explosion at Chernobyl, when the risk became a reality. This experience will forever color how the world looks at the various options for meeting energy needs.

Our choice now is either to rally behind the environmental revolution or to continue on the current path, moving toward a world where famine expands beyond the capacity of international relief agencies, where cancer reaches epidemic proportions and where the decline in living conditions now underway in some 40 countries continues to spread, dropping more and more of the world's poor below the survival level.

3 The Pivotal Role of Governments

Today there is a vast gap between what governments are doing and what they need to do for the environmental revolution to succeed. Most are taking some steps in the right direction, but even the most progressive countries are still in the early stages of the transformation.

Some governments are evaluating the policy instruments they can use to restructure the economy—tax policies, subsidies, regulations, research and development funding, procurement policies. Of all the instruments available, the partial replacement of income and other taxes with environmental taxes is by far the most effective tool to move the economy on to an environmentally sustainable path.

Taxing environmentally destructive activities, such as carbon emission, the generation of hazardous waste, and the use of virgin materials, permits the market to operate unimpaired, taking advantage of its inherent efficiencies while steering it in an environmentally sustainable direction. For example, taxing carbon emissions discourages the use of fossil fuels, while encouraging investments in energy efficiency and renewable sources of energy.

Environmental taxes address a principal weakness of market economies, namely that the costs of environmentally destructive economic activities are often not borne by those responsible for the damage. The costs of the respiratory illnesses of children in Mexico City are not borne by the motorists who generate the pollution. The costs of dealing with rising sea level in low-lying developing countries will not be borne by the industrial nations largely responsible for the carbon emissions that are driving climate change. As German environmental analyst Ernst von Weizsacher observes: 'Prices do not tell the ecological truth'. Environmental taxes provide a means of incorporating these indirect costs into the economy, correcting the shortcomings of the market.

Getting rid of subsidies that are ill-conceived or outdated can also speed the transformation. In the Soviet Union, where until recently energy has been supplied at prices far below those of the world market, there has been little incentive to use energy efficiently. As a result, more than twice as much is used to produce a dollar's worth of goods and services as in Japan, the most energy efficient industrial economy. Subsidies of pesticide use in developing countries have led to the excessive, sometimes reckless, use of farm chemicals. In the United States, subsidies such as those embodied in the federal policy of selling timber rights to logging companies at a fraction of cost encourage tree cutting at the expense of paper recycling. But today, from energy in the former Soviet Union to grazing rights under the new administration in the United States, some of those ill-conceived subsidies are coming to an end.

Carefully used, however, subsidies can accelerate the environmental revolution. For example, in the late 1970s California adopted a tax write-off to supplement the federal subsidy of investments in renewable energy resources. This, coupled with key regulatory changes that opened up the electricity market to generators other than electric utilities, set the stage for the state's emergence as a world leader in this field. California today generates more electricity from both wind farms and solar thermal power plants than does the rest of the world combined. This forward-looking state also leads in development of geothermal energy, accounting for roughly a third of the global total. Because of a bold governmental initiative taken over a decade ago, the world now has a window on the future, an early view of what an environmentally sustainable energy economy might look like.

Procurement policies are also an effective means of shifting priorities. For example, in government bureaucracies where the principal raw material is paper, a decision to buy only recycled products helps create a market, encouraging the industry to invest in mills that use recycled stock rather than freshly cut trees. The Clinton administration appears to be on the verge of such a decision. Similarly, replacing inefficient incandescent light bulbs in government offices with the new compact fluorescent ones that use only one fourth as much electricity would help create a viable market for the large-scale manufacture of mini-fluorescents while familiarizing both employees and the public with this energy saving technology.

On the waste management front, Denmark has emerged as a model by banning throw-away beverage containers, thus sharply reducing garbage generation. By forcing a shift to refillable containers, it has cut the energy invested in beverage containers by two thirds or more and lowered air and water pollution accordingly. Employment, meanwhile, may have increased, since reusing beverage containers is more labor intensive than manufacturing new ones.

The Netherlands has pioneered the use of bicycles for personal transportation. With a bicycle fleet of 12 million for a population of 15 million, this compact country has more than twice as many bicycles as cars. In Groningen, the largest city in the northern Netherlands, trips by bicycle already account for half of all trips.

These examples are drawn mostly from industrial countries, largely because Third World governments are too preoccupied with poverty, and particularly its spread over the past decade, to respond to global environmental threats. The challenge to national governments and the international community is to deal simultaneously with poverty and environmental degradation. The iron curtain is gone, the poverty curtain has become ever more visible. The ideological fault line that divided the world for nearly half a century has disappeared, only to be replaced by an economic fault line dividing the world between North and South.

While he was president of the World Bank, Barber Conable noted that 'as the 1990s unfold, success in reducing poverty should be "the" measurement of global economic progress'. It is difficult to imagine an effective international environmental effort if a strong net flow of capital from rich countries to poor ones is not restored. As others have noted, when we reach the point where half the world watches the other half starve to death on television, civilization will have come to an end.

If deepening poverty continues to threaten the survival of part of humanity in the short run, the environmental revolution will fail. People facing starvation today are not likely to worry about the effects of climate change tomorrow. We can no longer separate the future habitability of the planet from the current distribution of wealth.

The new reality distinguishes this decade from those that have gone before. It is not possible for part of the world to follow one path and the rest to follow another. No country can stabilize its climate in isolation. No country, acting unilaterally, can preserve the earth's biological diversity. If even one country continues to discharge large quantities of CFCs into the atmosphere, then people everywhere are doomed ultimately to rising levels of skin cancer, more frequent cataracts, the suppression of immune systems and the associated vulnerability to infectious diseases.

The poor have something the rich desperately need, namely their cooperation in preserving the global economy's environmental support systems. If affluent societies are unwilling to help the poor deal with life-threatening poverty, the developing world may be reluctant to devote much time or many resources to these threats. Although governments of rich nations such as the United States have in the past resisted it, a meaningful, sustainable development strategy anywhere must now embrace the satisfaction of basic human needs everywhere.

In effect, the environmental revolution depends on re-defining security. Robert McNamara—uniquely suited to comment on this as both former US Secretary of Defense and former President of the World Bank—argues that the time has come for a massive global reduction in military expenditures. Observing that Third World military expenditures totaled roughly $170 billion in 1988, nearly as much as was spent for health care and education combined, he urges that they be roughly cut in half over the remainder of this decade.

At this point, given the resources needed for the environmental revolution, national governments may want to examine the experience of Costa Rica, which abolished its army in 1949, leaving the maintenance of law and order to the police. As a result of this bold move, Costa Rica was able to spend more on education, health and family planning, and is now among Third World leaders in such basic social indicators as infant and maternal mortality, literacy and life expectancy. Indeed, life expectancy there is 77 years, two years more than in the United States.

Given the deteriorating state of the planet and all that portends, there is now a need for other similarly bold initiatives by national governments. Such actions are not a substitute for international cooperation, but they can provide a strong psychological boost for all of humanity.

4 Corporations Facing Change

Of all the sources of resistance to the environmental revolution, none is greater than that coming from corporate boardrooms. Accustomed to internalizing profits while externalizing costs, by passing on the expense of environmental destruction and disruption to society at large, corporations are not eager for wholesale reform. Although some companies are actively supporting the transformation, they are often younger firms, not those in the old-line industries. Unfortunately, too many companies assume that exploitation of the environment for dumping waste or acquiring raw materials is essential to their success, much as factory owners early in the industrial revolution thought their survival depended on the exploitation of child labor.

Corporations are facing an enormous amount of change in the years ahead, some far more than others. Much of the change will originate with governments, as public environmental concerns translate into environmental taxes, regulations and procurement policies. Some will come from environmentally sensitive consumers. And some will come from environmentally sensitive corporate leaders within the business community. Corporations will change in response to environmental threats, but will they do it with foresight, anticipating the adjustments needed? Or will they delay, possibly pushing society to the brink of catastrophe?

Since mid-century, corporate investment patterns have been shaped by the enormous worldwide growth in consumer purchasing power interacting with advancing technology and new products. Research and development on new products, combined with market research on what consumers needed or could be persuaded to buy, provided the grist for long-term corporate production planning. The fourfold expansion of the global economy over the last 40 years, coupled with the internationalization of markets, created a unique growth opportunity for enterprising firms. During the latter part of this four decade span, pollution control regulations and the banning of environmentally destructive products, such as CFCs or dangerous pesticides, began to affect business investments—but only in a minor way.

In the years ahead, environmental influences can be expected to increasingly dominate the evolution of the global economy. Phasing out fossil fuels, shifting to a reuse-recycle economy, protecting the stratospheric ozone layer, reducing air pollution and acid rain, minimizing hazardous waste generation and adapting to water scarcity are among the environmental influences shaping the economic future of corporations.

The economic restructuring in prospect is not trivial. The shift from fossil fuels to a solar-hydrogen energy system will affect every sector of the economy, from transportation to food. At present, transport systems are largely fueled by oil, either gasoline, diesel fuel or jet fuel. As the shift away from fossil fuels progresses, the use of electricity for transportation, now confined largely to rail systems, will likely become common for automobiles and trucks. And in locales distant from cheap electricity generating sites, hydrogen is likely to become the fuel of choice for cars, trucks and buses, as well as for airplanes everywhere. Efforts to reduce air pollution in California are leading to regulations on manufacturing that will require 2% of all cars sold in the state after 1998 to have zero emissions—that is, to be electric.

Efforts to put agriculture on a more sustainable footing will lead to more comprehensive nutrient recycling and integrated pest management strategies, affecting the market for fertilizers and pesticides. The food processing industry in industrial countries, now using more energy in processing, packaging and distributing food than farmers use in producing it, may face even greater changes.

Local limits on the earth's waste-absorptive capacity are forcing industries to redesign their manufacturing processes to generate less waste. Already doing this in response to government regulations, firms will find an even greater incentive to cut waste as environmental taxes are levied, such as those recently adopted on plastic bags in Italy and non-returnable beer and soft drink containers in Finland.

The oft-repeated notion that the business of business is business recognizes the obligation that corporations have to their stockholders, and implicitly the difficulties in unilaterally responding to environmental concerns that would put them at a competitive disadvantage. This short-term concern with profits contrasts sharply with the long-term nature of environmental concerns, setting up an inherent conflict in corporate management. Given the highly competitive setting in which firms operate, some corporate leaders welcome government regulation or environmental tax, as it permits them to reduce environmental damage without being put at a short-term competitive disadvantage.

Using their well developed commercial instincts, corporations attempt to exploit consumer environmental concerns in marketing their products, often confusing buyers with their advertising claims. The principal antidote to this practice is the emergence of so-called eco-labels, an approach pioneered in West Germany more than 10 years ago. The Blue Angel labeling program there has evaluated over 3,500 products in some 50 categories. In Canada, products bearing the Environmental Choice label first appeared on the shelves in 1989. Other European countries are modeling their labeling systems after that of Germany. The United States has two relatively new product evaluation groups, Green Cross and Green Seal, with the latter directed by a board consisting largely of leaders of environmental and consumer groups.

The premise underlying these programs is that environmentally sensitive consumers will alter their buying habits if given relevant information about products they routinely purchase. Decisions made by millions of consumers each day add up to major shifts in consumer demand. This, in turn, affects product design, method of manufacture and even whether some products appear at all. In Canada, where surveys indicate that 80% of the public is willing to pay 10% more for environmentally benign products, companies are obviously eager to exploit this concern by getting the seal on as many products as possible. Among other things, these programs evaluate the energy efficiency of household appliances and the recycled material content of products. In the case of Canada, for example, reusable diapers were given the seal of approval while throw-away diapers were not.

When public concern over environmental deterioration first surfaced, corporations often fought new regulations, arguing that industries and jobs would flee overseas. They typically reacted superficially, spending more on environmental advertising than on environmental action. These issues were, they believed, something to be dealt with by the public relations office, not by corporate planners or plant managers.

A number of firms have now moved beyond this and are beginning to assess their own operations in environmental terms, including such things as energy efficiency, paper recycling and hazardous waste generation. This is an important step, given the share of the global economy accounted for by the corporate sector. For example, in mid-1991 AT&T released an environmental strategy paper outlining near-term goals. This far-flung corporation, which employs some 275,000 workers worldwide, plans by the end of 1994 to phase out CFC use, to reduce waste from manufacturing processes by 25% and paper use by 15%, and to achieve a 35% paper recycling rate; by the end of 1995, it expects to cut toxic air emissions by 95%. Impressive though these goals are, the list is instructive as much for what it omits—goals for the reduction of carbon emissions and some hazardous wastes—as for what it includes.

At the next level, the challenge for companies is to ask whether there is a place for their products in an environmentally sustainable economy. For some, the changes in prospect are rather modest. Manufacturers of electrical household appliances, for instance, can concentrate mostly on making them more energy efficient and designing them so they can more easily be repaired or recycled. Companies now building gasoline-powered automobiles can shift to electric or hydrogen-powered cars without vast investments. Firms that produce incandescent light bulbs can make the switch to compact fluorescents simply by modifying their manufacturing facilities.

Many products manufactured on a relatively limited scale today have an enormous market potential in an environmentally sustainable world. Prominent among these are refillable beverage containers, compact fluorescent light bulbs, bicycles, photovoltaic cells, thermally efficient building materials, heat pumps, wind electric generators, high-speed rail cars, contraceptives, roof-top solar water heaters and water efficient plumbing appliances. Within the construction industry, the rapid growth areas will include building solar power plants, electrolysis plants to produce hydrogen fuel and commuter rail systems, and refurbishing and retro-fitting older buildings to boost their energy efficiency.

Another group of companies will have to take a more fundamental look at their products simply because there will be no place for them in an environmentally sustainable economy. For example, if climate stabilization is the goal, then coal and oil companies can either try to maintain the status quo and face a difficult, potentially bleak future or they can help develop renewable energy sources. Given the heavy damage from coal burning, some governments might one day even ban the use of coal, much as they have CFCs or nuclear power.

Some companies by their very nature contribute to carbon reductions. In California, several firms have raised over $2.5 billion in private capital markets for investment in wind turbine manufacture and wind farms, thus helping demonstrate that wind power is a viable energy source, one with a potential comparable to that of hydro-power, which now supplies one fifth of the world's electricity. Companies producing goods that directly or indirectly reduce carbon emissions, such as photovoltaic cells, bicycles and fax machines (which reduce the need for postal and private delivery services) are well positioned to help build an environmentally sustainable economy and to profit from it.

Like everyone else, corporations have a stake in a sustainable future. It is hard to sustain profits in a declining economy. Those within a given industry that see the need for change and move to the forefront will fare better than those who attempt to maintain the status quo. Current circumstances raise an intriguing question: are corporate interests and the survival of civilization now so intertwined that companies assuming a major environmental responsibility are likely to benefit economically, simply because they are responding to a deeply felt public concern about the future?

5 Mobilizing for Change

The environmental revolution by definition depends on an extraordinary amount of social change, a compression of history—squeezing centuries of change into decades.

Social change is not usually a smooth process. In the Soviet Union, democratization accelerated sharply under the leadership of Mikhail Gorbachev in 1985 in the form of glasnost. When the ill-fated coup d'état was attempted by the 'Committee of Eight' in 1991, it became the precipitating event that unleashed the forces of democracy and set the stage for dismantling the Soviet Empire.

Thus far, social change in response to environmental threats has also been a gradual process, slowly gaining momentum. Like democratization in the Soviet Union, it faces enormous obstacles in the form of people wedded to the status quo. Just as many Communist Party officials and bureaucrats resisted, there are many powerful vested interests in the existing world economic system, interests that are resisting the environmental revolution.

No one knows what the environmental equivalent of the Soviet coup d'état will be, the event that puts the environmental revolution on the fast track. It could be a crop-damaging rise in summer temperatures in the US grain belt that creates chaos in world grain markets, or a nuclear accident in France that exposes millions of Europeans to excessive radiation, forcing massive population relocations.

If the environmental revolution is to succeed, it will need the support of far more people than it now has. In addition to overcoming vested interests, it must also overcome human inertia. Up until now the environmental revolution has been viewed by society much like a sporting event—one where thousands of people sit in the stands watching, while only a handful are on the playing field actively attempting to influence the outcome of the contest. Success in this case depends on erasing the imaginary sidelines that separate spectators from participants so we can all get involved. Saving the planet is not a spectator sport.

I.3 Towards Environmentally Sound, Sustainable Development

E. Salim

Centre for Policy and Implementation Studies, Gedung Garuda Indonesia, 8-9 FL. JI. Medan Merdeka Selatan No 13, Jakarta 10110, Indonesia

Abstract

There must be new approaches and new concepts in development, especially in the developing countries, if global environmental destruction is to be halted and living standards of developing countries are to be improved.

If development is to be environmentally sound and sustainable, the technology and process of production should be considered for ways to minimize resources, energy, waste and pollution and capital per unit of output, and maximize labor absorption per unit of output at a humane wage level. This is especially crucial for developing countries.

The bottlenecks in achieving such development are caused by imperfections in the market mechanism and badly implemented government policies.

Global cooperation needs to be enhanced to overcome these bottlenecks and pave the way towards environmentally sound, sustainable development for developing countries in particular and the globe in general.

Dear Friends

Following Mr Lester Brown's explanation of the problem of environmental degradation, it is appropriate to concentrate on the question, 'What went wrong?'.

First, if we look at the conventional pattern of development, there has been a rapid increase since 1990 in gross world product, industrial output and energy use. This is a reflection of the dramatic increase in population.

The negative by-product, as explained by Mr Brown, is ozone depletion, acid rain and his 10 factors of environmental degradation. Additionally, the 1980s could be seen as a lost decade for Africa.

The two resulting questions are: how can we achieve development without environmental degradation, and how can developing countries receive their share of development?

According to the conventional model, economic development is measured by the growth in gross national product (GNP), the value of goods and services, and market prices. When pollution occurs, the cost of the services rendered forces up the GNP.

The GNP is derived from the production factor of land—including natural resources, labor, capital and entrepreneurs. The eco-system is not taken into account.

Further, this model is based on the assumption that people have unlimited wants and prefer more rather than less. Meeting this demand requires resources. The engine behind the demand for products is the increase in population growth and income.

Advocates of a free market economy expect private enterprise and competition to stimulate growth, whether the model is applied to a developing or developed country.

Population projections predict an increase from 5.2 billion in 1990 to 10 billion by 2050. This is based on the mortality rate decreasing at a greater rate than the fertility rate.

In developing countries the population will increase as health services improve and mortality rates drop. To meet the increase in population growth, the manufacturing output and gross world product must increase.

The question to consider is how to meet this demand. Should we follow the conventional pattern of development as outlined by Mr Brown, with its problematic consequences, or should we explore new types of development.

The proposal for the future is a new model called sustainable development, which is not limited solely to economic development.

It is modeled on a broader concept of development aiming to fulfill basic needs. The two items most relevant to developing countries are poverty alleviation and equal access to economic decision making. Of greater relevance to developed countries is the balancing of material consumption with consumption of non-materials such as culture, religion and education, which do not require natural resources.

On the equity issue, Indonesia's debt to Japan increased by 30%, not because of any new negotiations, but because the value of the yen appreciated. In order to increase its exports to pay the unexpected increase in debt, Indonesia has to exploit more natural resources.

Social issues, neglected by the conventional model, are given emphasis in the sustainable model. People do not live by bread alone, they need goals such as social justice and gender equity, a commitment to cultural, religious and racial diversity, and a move towards a stable population through reduced birth and mortality rates.

In political terms, the sustainable model supports political stability, participation in government by the people, transparent policy making and equal access to political decision making.

If sustainable development is to be achieved the political dimension must be addressed. Africa provides an example of its importance, for though the continent receives the highest portion of international aid, which is close to 8% of its gross domestic product, the 1980s can be regarded as a lost decade.

Ecological considerations and those concerned with the eco-system are vital. The eco-system has its own life support system which must be maintained. A process of development is needed that assures bio-diversity and preserves resources. We must consider ways of maintaining development without exceeding what resources can sustain.

One role of the eco-system is to absorb pollution. The question to be asked is how can we raise environmental resilience to ensure this happens? Enhancing the quality of the environment is part of the sustainable development model for the future. We know that this can be done. Japan has shown that it is possible to clean up rivers and get fish coming back. It is also possible to replant critical land.

Sustainable development is consequently much broader than economic development as it also takes social, political and environmental aspects into account.

There are several factors relevant to achieving the goals of a sustainable development model. Human resources are an important developmental factor. These can be viewed from both a physical and non-physical perspective. Rather than seeing such resources solely as units of labor for production, this alternative approach also considers the non-physical aspects of human resources and how these can be enhanced through training, and through better opportunities in education, health and cultural development.

In the process of achieving sustainable development, social capital is important. This covers family and community, as well as political parties and enterprises, and other organisations. Manufactured capital on the other hand comprises capital goods, machinery and technology. Ecological capital is natural resources and eco-systems.

Human resources, and social, manufactured and ecological capital are the four developmental factors which make up the objective function of the model. This means that economic, social, political and ecological developments all have a part to play.

The developmental functions are currently subject to certain constraints. Technology needs to be improved to overcome these difficulties. We have a huge unemployment rate in the developing countries and ways need to be found to maximize labor absorption. Organisational costs, capital input, resource efficiency and what Lester Brown called 'energy efficiency', expenditure and waste all need to be minimized per unit of output, while recycling per unit of output needs to be maximized. These are the criteria to choose in the process of production and technology, in order to achieve the desired development goals.

Given that the model is a theoretical one constructed along linear programed lines, with an objective goal, identified factors and areas of constraint, the question to consider is how best to proceed. The proposal is to make use of monetary means to achieve the goals of sustainable development. Monetary mechanisms such as incentive systems, environmental taxes, tradeable pollution permits, resource depletion charges, common resource pricing and internalizing externalities can all be used. These mechanisms have been implemented by Japan and the United States. They should now be implemented globally.

There are also non-monetary mechanisms which can be supported by law. Indonesia has environmental protection regulations but, as in most developing countries, the difficulty is the enforcement of the law. Education, information and transparent policy making are needed to implement non-monetary mechanisms. Introducing and enforcing rights to property and resources, alongside social control and participation in government by the people, are also important.

Monetary and non-monetary methods should both be followed. Governments have a responsibility to develop, as far as possible, a policy of internalizing the externalities, thus creating a market price structure that includes environmental costs. The market economy means that market prices are important but the danger is that they fail to include the environmental costs. Government intervention is needed to correct these market failures.

Developing a climate conducive to environmentally sound, sustainable development remains the most important priority. One of the issues to be considered is that of inter-generational equity. The needs and preferences of the current generation are reflected in the present political system and market economy. But the unborn babies, the future generation of the year 2000 and beyond, have no political voice, nobody to articulate their scale of preferences. To address this, in economic terms, means government intervention so that resources can be preserved. The political system must look towards the future. Special consideration should be given to protect the rights of future generations and prevent the present generation from exploiting resources for their own benefit.

Associated with these concerns is the issue of long-term planning. Economic issues are predominantly short-term, for as the famous economist Lord Keynes mentioned, 'in the long run we are all dead'. Because of this, it is important to work with long-term perspective planning using frameworks of 25, 10, or five years. The Japanese economy is the result of perspective planning. An example of this, brought to my attention by sensei Sabura Okita, was the sub-program policy of doubling income in the 1960s and 1970s.

Impact analysis is another area which needs special consideration. The impact of development on the environment, not reflected by pricing mechanisms, is (as previously stated) commonly referred to as 'externalities'. Government intervention is necessary to evaluate what this impact is, using methodologies and impact analysis. Mr Brown is concerned with the issues relating to energy. Energy poses problems because prices charged for it do not include such environmental considerations as pollution, CO_2 emission and depletion of oil stock. Consequently fossil fuel prices are artificially low. How can renewable resources compete with such low prices? Research into renewable resources appears unattractive because the advantages seem limited owing to the distorted price structure of non-renewable resources. It is proposed that work should be attempted on so-called shadow pricing so that the feasibility of using renewable energy sources can be realistically compared with the full environmental costs of fossil fuels.

Indonesia is now program planning its next 25 year plan for the years 1994-2020 and improved energy efficiency is an important goal. The question as to whether renewable sources of energy are a viable proposition in the next 10 or 15 years is confronting everyone concerned with environmental issues. We are considering it either as a source of supply for the national grid system or for industrial development. I have been advised by experts in Indonesia that renewable sources of energy are feasible for agricultural development and for water heating but not for a national grid system or industrial development. The exploration of the potential of renewable energy is of crucial importance, specifically for Asia. There is a current concern in Indonesia that if China moves into industrialization, using coal as a major source of energy, acid rain, brought to the region by pervading winds, could be a potential by-product. I support Lester Brown's proposal that for the next 25 years Japan concentrates on pioneering a renewable source of solar, wind or wave energy. At present it seems unrealistic to expect results in less than that time.

If we follow a development model that internalizes externalities so that environmental costs are included, prices will rise. When, as Minister of the Environment, I raise my hand in cabinet meetings, to ask that these costs be covered, the reaction of my colleagues is to complain that the cost will go up. In order to protect the poor we need social safety nets. When a price appropriate for environmental management is developed it is the poor who will suffer. If the internalizing environmental costs are added to that of fossil fuel, for instance, kerosene diesel prices will rise and the poor who are consumers of kerosene will be hit.

Reassessments of the methods by which the GNP is calculated need to be considered. Efforts are currently being made to include assessment of the depletion of resource stock into the national product. Similarly work on daily income is underway to develop an index of sustainable economic welfare. If sustainable development is a broad concept which covers economic, social, political and ecological areas, then its indicators need to be broader than those of the GNP alone.

Having outlined what sustainable development is and the road to reach it, we need to give attention to how this can be achieved through global cooperation. Following the end of the ideological conflict between capitalism and communism, the capitalist market-oriented economy is now global and its tenets preached throughout the world. Why is there this emphasis on materialistic development? The answer is that it can be measured. Attention is concentrated on seven major countries in the world (the 'group of seven' or G7) which dictate the political and economic development of all other countries. Trade, aid and economic activities are now conditional on democracy, human rights and environmental considerations. The economic and political power of the G7 determines the rate and general agreements on trade and tariffs. Developing countries experience the frustration of having to accept unsatisfactory decisions because they lack the concentrated economic and political power of the G7. The fall of the United States dollar in relation to the Japanese yen has had the result of increasing Indonesia's debt to Japan. Every point that it declines increases our debt repayment.

Consideration should be given to the impact of the increased share of the market by multinational corporations, which sometimes exceeds the gross domestic product of individual countries. In 1989 Toyota's turnover of US$60 billion exceeded that of the gross domestic product of the former Yugoslavia during the same period. Other large companies acting in concert can exercise considerable economic influence. With the increasing power of the capitalist economy they are now telling us how and what to produce and consume.

Regional cooperation also needs to be examined. The European Community and the North American Free Trade Area are regional blocks operating predominantly among developed countries. These groups, using preferential trading agreements, protectionism and the imposition of conditionalities, contribute to economic discrepancy between North and South.

The possible decline of per capita income in the developing countries and the increase in the per capita income of developed countries, such as Japan, was mentioned by Lester Brown. If this happens, if this is the future of our global economy, how are we to reach sustainable development in the world?

Ladies and gentlemen, once the concept of sustainable global economy is clear and the means and policy understood, it is important to work for the renewal of global partnerships. The goal of such a partnership is the forging of a balance between material and non-material development. In an important article by Huntington in 1993, the author argues that the end of the East-West conflict between communism and capitalism will be followed by a new struggle between the West and the non-West. The civilization of the non-West in the Asia-Pacific area is based on Confucianism and Islam. Therefore, according to this article, there will in the future be a clash of civilization between the West and Confucianism and Islam. Such a thesis fails to convince me, because the issue is not about confrontations between ideologies and religions, but about types of development. And the type of development which is not environmentally friendly is one which too heavily emphasizes materialistic concerns.

The promotion of non-material development is what is needed and within that context religion and culture become important. It is for these reasons that I deliberately include social considerations as a part of sustainable development.

The international movement towards materialistic development must be prevented, otherwise natural resources will be exhausted and the environment degraded.

Representatives of major religious groups meeting in Chicago recently voiced similar views. Pope John Paul II in his letter of April 1991, *Dererum Novarum*, expressed the concern that the domination of the capitalist economy tends to make the world too materialistic. Sustainable development offers a major solution in redressing the balance between material and non-material types of development.

Along with this solution goes the need to disperse economic and political power by restructuring and strengthening the United Nations. In 1995 the United Nations will be 50 years old. There have been substantial international changes during that time. Japan and Germany have become world economic powers and both have seats on the Security Council.

The future of nations is often determined by the G7 which is not accountable to the United Nations and can completely disregard it. The question to consider is the appropriateness of international issues being decided by a forum in which the concerns of the United Nations are ignored.

Increasing multilateral, bilateral and private cooperation between North and South through trade flow is of utmost importance. Leaders of developing countries don't want aid, they want to earn development through trade. Current trade barriers and quota systems are hampering this growth. In Indonesia the production of textiles is restricted because of trade barriers and import quotas.

Investment flow and the transfer of technology are areas that must be addressed. Important gene resources exist in the South and these need to be safeguarded and conserved. An example is the gene resources found in the tropical forests of Sumatra, Galimanda and Indonesia used in pharmaceutical products for cosmetics and industry. Technology from the North is needed to transform these resources. Intellectual property rights would assure those in the North of satisfactory returns for their research.

We in the South receive no compensation for conserving our environmental resources though we are told by the North to conserve, and not to exploit our tropical forests. When my people ask what we get in return, I explain that it is important to prevent global climatic changes due to global warming. In response they say that as climatic changes are important for the bio-diversity of the whole earth, the international community has an obligation to make recompense to the people whose forests are conserved. Transfers of technology, such as in the gene resources example, offer ways this can be done while assisting people in their efforts to manage the environment.

Debt relief is a crucial issue for developing countries. Seventeen of the least developed countries have a debt surface ratio of over 50% of export earning, so when they export, 50% of what is earned is lost in debt repayments. How can such a country develop? The rate of debt relief is so high that, though aid is being received, it is less than the outflow which is being paid to developed countries to service the debt. Fresh aid is then needed. I am also aware, whenever

I mention this issue, that there is a feeling of fatigue, as people are tired of talking about aid.

So what must be done to achieve global cooperation? Trade flow and investment flow must be enhanced. Technology transfers must be made possible and human resources built on. Appropriate debt relief and balanced aid flows must be available for developing countries. Such steps will enable developing countries to implement sustainable development. A broad focus for action is essential to prevent serious environmental degradation. The potential seriousness of this degradation is apparent from the development patterns of the developed countries, a process which we in the South are now following.

We know that there is the opportunity for implementing a different concept of development. We cannot follow the way Japan has moved with conventional development or the path of conventional development espoused by South Korea, Taiwan and China. We want to move towards sustainable development which is a different kind of development. To do this, resources are needed. We seek a fair trade relationship without tariff barriers. We ask for fair compensation for technology transfer, fair investment flow and debt relief. Given these resources we can implement sustainable development.

The population of 10 billion predicted by the year 2050 is expected to be concentrated largely (80%) in developing countries. Given these projections, development could cause environmental destruction. The concept of sustainable development will prevent this catastrophe. I am happy that this second IBM symposium is discussing this issue in an atmosphere which facilitates dialog.

Let us put the problem squarely on the table. I am not a diplomat and I say black is black. Please forgive me if I have talked too frankly but I see an end to the tunnel; the light at the end of the tunnel is there. It is possible, it is feasible. What is needed now is the transformation of world politics and the globe, which can be achieved through international cooperation.

I.4 Technological Breakthrough and Global Cooperation

T. Kashiwagi

Professor, Faculty of Engineering, Tokyo University of A & T, Japan

Abstract

In this paper, technologies for clean energy conversion as well as concepts for rationalized clean energy utilization and conservation are discussed, and an outline is given of the Cooperative Research and Developement (R & D) Program on Appropriate Technologies proposed by the Japanese Agency of Industrial Science and Technology.

In Section 1, an approach is made to environmental problems from the viewpoint of climatic changes, and the importance of the time factor in these changes is discussed. Although severe climatic changes such as glaciations have occurred in the past, they took place over periods lasting from tens of thousands to 100,000 years. On the other hand, the current changes in global climate can be related to the industrialization process, which started less than 200 years ago.

Section 2 consists of a discussion on environment-friendly clean energy and technological developments. The main established clean energy technologies are listed, followed by a discussion of three future energy technologies which through their development and diffusion are expected to generate significant breakthroughs—namely nuclear fusion, space solar power and energy storage utilizing superconducting materials.

Section 3 presents an overview of Japan's large, national clean energy projects, the Sunshine Project and the Moonlight Project, which have recently been combined into the New Sunshine Program. An outline is also given of the Cooperative R & D Program on Appropriate Technologies, which is one of the three core programs of the New Sunshine Program.

Section 4 discusses the need for a systematic approach to energy conservation by means of cascaded energy utilization, co-generation and unexploited energy utilization systems. The concept of cascaded energy utilization systems is

be compared with their own pre-industrial levels, freons, which were invented in the United States in the 1930s, have been extensively used during the past 40 years and hence account for many unknowns in terms of climate change prevention planning for the future.

In particular, the greenhouse effect of chlorofluorocarbon (CFC) is 10,000 times greater than that of CO_2 gas, which means that the effect of even low freon concentrations cannot be overlooked. In view of the ozone layer depletion problem, an international consensus has been reached toward complete abolition of CFC use by the year 1995. The problem remains, however, that the life span of restricted freons is extremely long in comparison to that of CO_2, as shown in Table 1. This is why plans for curbing global climatic changes must be devised for a 50 to 100 year period, which by engineering standards is considered a long-range time span.

2 Environment-friendly Clean Energy and Technological Developments

2.1 Established clean energy technologies

The fragility of the world energy supply structure plus the variety of factors requiring consideration when selecting energy alternatives make it difficult to conceive a solution in terms of one single energy source. At present, a *best mix* of diversified energy sources is often mentioned as a good solution. In the following paragraphs, a broad outline will be given of clean energy sources and technological strategies with a view toward rationalized energy utilization and global warming prevention.

Global warming, as is well known, results from the rise in atmospheric temperature due to increased concentrations of greenhouse gases, which absorb infra-red radiation emitted from the ground. Since about 50% of the greenhouse gases consist of CO_2, and about 80% of the total amount of CO_2 emitted comes from the combustion of fossil fuels, namely oil, coal and natural gas, there is clearly a very close relation between the formulation of global warming prevention strategies and the choice of energy alternatives as well as the rationalization of energy use. A number of clean energy alternatives and related technologies for global warming prevention are summarized as follows:

[1] CO_2 absorbing energy sources

- bio-energy:
  biomass utilization
  biomass conversion
- ocean energy:
  power generation by temperature differentials
  waves and tidal currents

[2] Non-CO_2 emitting energy sources

- nuclear energy:
  technologies for safety

- solar energy:
  photovoltaic power generation
  thermal power generation
  active and passive solar systems
  solar chemistry
  solar power generation in space
- geothermal energy (power generation):
  flash
  total flow
  binary cycle
  hot dry rock
  magma
- wind energy (power generation):
  new type windmills
  combined wind energy systems
- hydrogen energy:
  high efficiency hydrogen production from water
  energy transport and storage using hydrogen alloys
  hydrogen turbines

[3] Technologies for reducing CO_2 emission
- high efficiency energy utilization systems (Figure 3f):
  integrated energy systems—biomass, synthetic fuels and hydrogen (Figure 4)
  unexploited energy (Figure 5)
  heat cascading (Figure 6, Figure 3c)
  co-generation (Figure 6b, Table 2)
  resource recycling
- fuel conversion technologies:
  coal gasification and liquefaction
  natural gas liquefaction and decarbonization

[4] CO_2 fixation technologies
- solar bioreactors (Figure 7)
- methanol generation by direct hydrogenation (Figure 8)

2.2 Future clean energy technologies

Human society in the twenty-first century will have to resort to technology and information in order to confront in earnest a number of rapidly emerging problems of civilization, such as global environment deterioration, population growth, lack of resources and energy, and food shortages.

Since energy and environment measures involve high grade and complex system technologies, it is usual for their implementation to require long lead times, which makes it indispensable to provide long-term continuity to research and development in response to their needs. Moreover, the need for substantial capital investment calls for the realization of effective fund allocations. Hence, the development of sustainable energy and environment technologies for the twenty-first century requires strategic development aimed at practical implementation, with systematic and comprehensive evaluation.

This sub-section will discuss three energy technologies which through their development and diffusion are expected to generate significant breakthroughs in the future: nuclear fusion, space solar power and energy storage utilizing superconducting materials.

2.2.1 Nuclear fusion

Nuclear fusion power has all the characteristics of an ideal power generation system for the twenty-first century. Since the success of nuclear fusion will mean humankind having access to an inexhaustible and environmentally harmless energy source, it is necessary to plan bolder research and faster development realizations.

There are two major ways of achieving power generation by nuclear fusion, namely magnetically confined nuclear fusion (steady-state reactor)—where ultra-high temperature plasma is confined by a magnetic flux container—and inertial nuclear fusion (pulsed reactor)—where a high power laser strikes heavy hydrogen pellets thus achieving instantaneous nuclear fusion.

For the Tokamak reactor, on which most of the R & D in the past 30 years has been done, records of energy input to output ratio equal to one have been obtained with a critical plasma experimental device (JT-60). The achievement of zero output conditions is also expected to be possible in the next few years. Therefore, it may be said that nuclear fusion demonstration on a scientific level has been completed, and that an engineering demonstration stage targeting actual size reactors has been started.

Coming R & D on nuclear fusion should be carried out according to three steps:

- experimental reactor: achievement of auto-ignition, realization of long-sustained combustion and establishment of the fundamentals of reactor engineering technology;
- actual size reactor: realization of steady-state core plasma and technical demonstration of power generation at plant scale;
- demonstration reactor: verification of power plant economics.

Regarding the device for use in the experimental reactor, the Tokamak type is expected to yield results in the shortest time, since among magnetically confined type reactors it features the largest amount of accumulated data on core plasma technology and also the best plasma properties. R & D on the Tokamak type is also recommended for the actual size reactor stage.

The expected time schedule for realization of the steps outlined above, would be, in a best case scenario, the experimental reactor starting operation around the year 2005 and the actual size reactor commencing in the 2020s. As a result of these R & D activities, the practical use of nuclear fusion as an energy source may be expected for the mid-twenty-first century (around the years 2030–2040).

Nuclear fusion for power generation requires very large reactors, thus involving extremely high R & D costs. However, if their characteristics as ideal inexhaustible energy sources for the twenty-first century are considered, it may be in the interest of countries with limited energy resources to actively engage in R & D activities, both independently and as participants of the International Tokamak Experimental Reactor (ITER).

2.2.2 *Space solar power*

The main space technology which Japan is expected to have established independently in the year 2000 is the capability to launch 2,000 kg disposable rockets (H-II) into geostationary orbit. Japan is also participating in the international Space Station Project with the Japan Experimental Module. Training of astronauts to travel in the module is done in cooperation with NASA.

The space solar power system under consideration has three subsystems: the space solar power satellite, the equipment for microwave power transmission to earth, and the power reception facilities on earth (rectenna sites). In order to be economically viable, the project must be of four or five million kilowatt (kW) class, with several units generating 1,000,000 kW each. This makes it difficult for only one country to independently undertake development of the extensive technology required by such a large-scale power generation system. Hence, the promotion of an international project, from the conceptual design and elemental technology development stages, is recommended.

The time scale envisaged for the space solar power project is as follows. System studies, development of elemental technologies, demonstration of microwave power transmission technologies and studies on the ecological impact of rectenna sites are to be carried out by the year 2000. A small-scale demonstration solar power generation unit (50,000 kW class) is to be launched around the year 2010, assembled in orbit and operated to demonstrate the feasibility of space solar power. The practical utilization of a 1,000,000 kW unit is to start around the years 2040–2050. The space power satellite, to remain in geostationary orbit, will not suffer structural degradation and may thus be considered as semi-permanent. In-orbit maintenance and replacement is envisaged for aged parts. These developments are aimed at bringing space solar power to account for 20–30% of the Japanese energy supply around the year 2100.

2.2.3 *Energy storage with superconductors*

The development of practical utilization technologies for superconductors is now being strongly promoted. One superconductivity application expected to be introduced after the year 2000 is the superconductor power generator, of which 10,000,000 kW class units are to be introduced to nuclear and thermal power generation plants.

The use of energy storage with superconductors is expected to improve the electric energy storage efficiency from the 70% currently possible with pumping water to above 90%. The proportion of nuclear power generated for power supply is expected to substantially increase in the years 2030–2050, and the practical utilization of large-capacity energy storage with superconductor technology is thus needed for its high effectiveness in improving both system stability and voltage stability.

If small-capacity units (less than two megawatt hours) of energy storage with superconductors for system stability control can be put into practical application from mid-capacity superconductor units for supply with rapid load fluctuations, the practical utilization of large-capacity superconductors (five gigawatt hour class) for daytime peak-shaving can be envisaged for around the year 2040.

If application of high temperature (ambient temperature level) superconductivity becomes possible, construction costs may be reduced by 15–20% and the economics will improve substantially.

3 Comprehensive Development of Clean Energy: Japan and its Contribution to the International Community

Energy and environmental problems in the twenty-first century will inevitably have a global character, which means that energy and environment measures developed to meet the needs of specific countries, such as Japan, will be of use and necessity to other countries as well. If the related technologies are thought of as international public property, it becomes necessary to establish an open door international cooperation system from the R & D stage, and to actively promote technological transfer so as to broaden the dissemination of R & D results. Such international cooperation will make it possible to tackle global environmental problems on a worldwide level and ensure mutual benefits to all countries involved.

3.1 Clean energy large national projects in Japan

Japan now accounts for 5% of the total world energy consumption and, next to the United States, is the world's second largest user of energy. A plentiful, inexpensive supply of petroleum enabled Japan's consumption of energy to increase swiftly during the early 1960s. Since then, however, all of this petroleum has been imported and, as demand increased, self-sufficiency in energy fell from 76% in 1955 to only 19% today.

With this background, the Sunshine Project was initiated in 1974, with the aims of securing solutions to basic energy problems and of harmonizing energy utilization in order to protect the environment. This project is the first large-scale, long-term technological development effort which deserves to be designated as a national project. It has as its object all aspects of new energy technology, with the exception of nuclear power technology.

Major development targets have been set in four fields:

- solar energy,
- geothermal energy,
- gasification and liquefaction of coal,
- hydrogen energy.

R & D also includes alternative energy sources such as wind, ocean, etc.

Another large national project, nicknamed the Moonlight Project, was launched in 1978 as an R & D program for developing advanced energy conservation technology, now being carried out by the Ministry of International Trade and Industry's Agency of Industrial Science and Technology. Under the Moonlight Project, R & D on energy conservation is conducted systematically by national laboratories, private firms and universities working in cooperation.

The name Moonlight was chosen because energy conservation technology should enable the efficient use of even small amounts of energy, as the moon shines by reflecting light from the sun.

Recently, five large-scale projects have been carried out:

- Advanced Battery Electric Power Storage System,
- Fuel Cell Power Generation Technology,
- Super Heat Pump and Energy Accumulation System,
- Superconducting Technology for Electric Power Apparatuses,
- Ceramic Gas Turbine Project.

The R & D Project for Environmental Technology includes studies on CO_2 fixation by artificial photosynthesis, separation of CO_2 and biodegradation of chemical compounds. These projects, which have been carried out in parallel, have lead to the accumulation of a substantial technological stock for effectively coping with energy constraints.

The global environmental consequences of energy use are now calling for a comprehensive approach in paving the way for sustainable economic growth by overcoming energy and environmental constraints simultaneously. In response to this urgent need, the Ministry of International Trade and Industry has established the New Sunshine Program (R & D Program on Energy and Environmental Technologies) by integrating the Sunshine Project, the Moonlight Project and the Global Environmental Technology Program. Through the integration of these R & D activities, effective and accelerated achievement of R & D in the field of energy and environmental technologies is expected by means of co-utilization and supplementation of various technologies common to new energy, energy conservation and environmental protection.

3.2 The New Sunshine Program

The New Sunshine Program comprises three technological systems:

[1] Innovative technology development: the development of innovative energy and environmental technology is accelerated to implement the Global Warming Prevention Action Plan, while focusing on important subjects in the course of progress.

[2] International, large-scale collaborative research: an international, large-scale collaborative research program is prompted to implement the Earth Regeneration Program 'New Earth 21'.

[3] Collaborative R & D on Appropriate Technology: a collaborative research and accelerated development program is promoted for fitting energy and environmental technology to the conditions of the partner country to ease

energy restrictions and address environmental concerns in neighboring developing countries.

Figure 9 illustrates the concept of the New Sunshine Program. The development of innovative technology will contribute to the Global Warming Prevention Action Plan of Japan, while the collaborative R & D on appropriate technology will contribute to innovation in neighboring developing countries. The results of these two projects as well as international, large-scale collaborative research will contribute to the Earth Regeneration Program.

Figure 10 shows the contribution of the New Sunshine Program to the Earth Regeneration Program, while Figure 11 shows the increasing contribution of the New Sunshine Program to the domestic energy supply and to the reduction of CO_2 emissions.

Among the many subjects in the new energy and energy conservation technologies that have been studied in the Sunshine Project and the Moonlight Project, solar cells and fuel cells show a 'positive circulation', or favorable synergistic effect. The development of technology for these led to an increase in demand and a reduction in cost, which was triggered by recent major technological innovations. Nevertheless, full-scale introduction of the technology still requires lead time to establish sufficient production capacity and develop adequate marketing. Therefore, R & D of solar cells and fuel cells must be accelerated to maximize their supply potential and realize the supply of energy alternatives to oil for the Global Warming Prevention Action Plan.

Figure 12 shows a cost reduction scheme for these two technologies. Accelerated promotion of the New Sunshine Program will reduce the cost of electricity from both technologies to 20 yen per kilowatt hour by the beginning of the year 2000. Innovative technologies are expected to contribute markedly to the solution of energy and environmental restrictions in the medium- to long-term.

Three innovative technological subjects were initiated with the New Sunshine Program in addition to the continued, accelerated implementation of R & D of subjects such as solar cells and fuel cells from the Sunshine Project and the Moonlight Project. These new subjects were derived form the results of the Sunshine Project and the Moonlight Project and are expected to yield significant medium- to long-term benefits:

- broad area energy utilization network system technology (eco-energy city system),
- international clean energy network using hydrogen conversion (World Energy Network: WE-NET),
- economically and environmentally compatible combustion system (lean burn de-NOx catalyst system).

Coal is expected to show a drastic worldwide increase in demand. The New Sunshine Project deals with coal conversion technology in response to the increasing request for its use under environmentally friendly conditions. Coal conversion technology that has long been under study is integrated with hydrogen production technology using renewable energy and lean burn de-

NOx technology to develop the Environmentally Friendly and Economically Viable Coal Conversion Complex Technology. These technological development issues are outlined below.

- Broad area energy utilization network system technology (eco-energy city system), Figure 13

Outline: Advanced cascaded and combined thermal energy recirculation system based on innovative technology that recovers waste heat from facilities and transports the recovered energy efficiently to remote urban consumption areas.

Breakthroughs: [1] Key technology developments: heat recovery technology with a maximum utilization of sensible heat and with temperature lifting technology; heat transport and storage technology using chemical reactions; multi-functional heat supply technology. [2] Structuring and optimizing the total system by matching multiple and wide-area heat sources to the energy demand.

Expected results: When unexploited energy in coastal industrial regions near major urban areas is used to the maximum extent, Japanese energy consumption can be reduced by 6% and CO_2 emissions can be reduced by 9%.

- International clean energy network using hydrogen conversion (World Energy Network: WE-NET), Figure 14

Outline: World scale clean energy system based on innovative technology from production and transfer to consumption using hydrogen as the medium.

Breakthroughs: [1] Development of elemental technologies relating to hydrogen utilization: high efficiency and large-scale hydrogen production technology; high density hydrogen storage and transfer technology; hydrogen turbine technology. [2] Structuring and optimizing world-scale systems related to production, distribution and utilization.

Expected results: When fully implemented, the system will contribute to the reduction of worldwide CO_2 emissions by 10% in 2030 and 20% in 2050.

- Economically and environmentally compatible combustion technology (lean burn de-NOx catalyst technology), Figure 15

Outline: Exhaust gas de-NOx catalyst technology applicable to diesel engines and lean burn engines aiming at the improvement of vehicle fuel costs and the simultaneous reduction of CO_2 and NOx emissions.

Breakthroughs: Development of elemental technologies related to new catalysts: analysis of material and structure, heat resistance, corrosion resistance, and improvement of catalysts' life in atmospheres with high oxygen concentration.

Expected results: Compared with conventional systems, the technology contributes to reducing Japanese energy consumption by 2%, CO_2 emissions by 2% and NOx emissions by 40%.

- Environmentally Friendly and Economically Viable Coal Conversion Complex Technology

Outline: Coal liquefaction complex technology aiming at drastic reductions of product cost and CO_2 emissions by combining a hydrogen production system employing renewable energy and a de-NOx lean burn combustion system.

Breakthroughs: [1] Establishment of key technologies related to coal liquefaction. [2] Integration of hydrogen production and lean burn catalyst technologies with coal liquefaction technology.

Expected results: Reduction of CO_2 emissions by 2%; reduction of product cost from $35/barrel (bbl) to $25/bbl.

3.3 Collaborative R & D Program on Appropriate Technologies

The Japanese Agency of Industrial Science and Technology proposed the Collaborative R & D Program on Appropriate Technologies as one of the three core programs of the New Sunshine Program starting in the 1993 financial year. The underlying concept of the program is recognition of the necessity and importance of international collaboration for reducing constraints on solving energy and environmental problems on a global scale. In particular, developing countries often face technological barriers to adequately address energy and environmental issues. In this respect, industrialized countries should actively support the self-help effort of developing countries to further advance global problem solving activities on these issues.

Japan has extended technological support for energy and the environment in developing countries. A typical example is the Green Aid Plan, a part of Official Development Assistance (ODA) in which the Ministry of International Trade and Industry actively promotes a wide variety of support activities. These activities include cooperation for state surveys in developing countries on problems of energy and the environment, cooperation in developing human resources through sending specialists to these countries and the acceptance of trainees into Japan for training technical personnel, and cooperation in R & D activities, such as fundamental research (Institute for Transfer of Industrial Technology (ITIT) project), pilot plant studies and plant validation operations. In addition, cooperative research projects that were derived from the Sunshine Project and the Moonlight Project in advanced technology fields for new energy and energy conservation have been effective.

The New Energy and Industrial Technology Development Organization will administer the Collaborative R & D Program on Appropriate Technologies. The organization conducted a survey on new energy and the promotion of energy conservation technology (ASEAN Caravan) in South-East Asian countries (Thailand, Indonesia, Malaysia and the Philippines) from February 1990 to March 1992. According to the survey results and to requests from these countries, cooperative research on the following topics is expected by the countries surveyed.

[1] Photovoltaic power generation technology:
- village power supply system
- isolated island power supply system
- battery charger station power source
- water pump power source
- other photovoltaic power generation technologies

[2] Solar heat utilization technology:
- agricultural products drying technology
- fixed temperature warehouse and refrigeration technology

[3] Geothermal power generation technology:
- geothermal exploration and development technology
- multi-purpose hot water utilization technology
- silica scale removal technology

[4] Wind power generation technology:
- isolated island and outlying region electrification system
- general wind power generation technology

[5] Fuel cell power generation technology:
- small-scale local distributed power generation system

[6] Energy conservation technology:
- factory energy conservation technology
- efficient oil use technology
- technology for highly efficient combustion of poor quality oil

[7] Technology for energy recovery from waste, etc.:
- technology for power generation from municipal waste
- biomass energy technology (use of agricultural wastes from rubber, palm oil, cocoa etc.)
- technology for effective use of peat
- technology for methane gas production from domestic animal excretion
- technology for alcohol production from tapioca (cassava starch)
- technology for comprehensive utilization of eucalyptus components

Some of these subjects are established in Japan, while others are already included in the cooperative research program. Of these, photovoltaic power generation technology, geothermal exploration and power generation technology, wind power generation technology and coal liquefaction technology are either in the R & D stage in the Sunshine Project and Moonlight Project, or they are at a stage where commercialization appears attractive. Subjects in both of these stages have different characteristics to those of conventional technological cooperation.

Specifically, further investigation in these subjects is centered on the natural environment, social conditions (including the style for using the technology concerned), the technological capabilities of the partner country to implement practical applications and the need to modify the technology to match the conditions in the partner country, as required. Thus, these subjects are accompanied with certain risks. Accordingly, these cooperative research subjects are best suited to the national institutes, firms that have sufficient experience in the Sunshine Project or Moonlight Project, and firms that can receive instruction from these experienced institutes or private firms. These projects, which cover the range from basic studies to practical applications, can provide the partner country with extensive know-how.

The main items in the Collaborative R & D Program on Appropriate Technologies, proposed as a part of the New Sunshine Program, are summarized below.

[1] The basic elemental and system technologies of the target collaborative subjects are established either in the Sunshine Project or the Moonlight Project.

[2] The scope of collaborative R & D is concentrated on matching the technology developed in [1] to the partner country's conditions, such as the natural environment, social conditions (ie style of using the technology), and technological capability.

[3] The Collaborative R & D Program deals with all stages, from feasibility study to pilot plant design, fabrication, installation, operation, evaluation and trouble-shooting.

[4] Education and training of the personnel necessary for collaborative R & D are conducted both in Japan and the partner country, as needed.

[5] A long-term technology support system is established to fully integrate the technology.

[6] In addition, the creation of new businesses in the future using the collaborative technology is considered to ensure that the technology is firmly settled in the partner country by its own effort.

[7] The participants of the Collaborative R & D Program are limited to those national research institutes and private firms that have experience in either the Sunshine Project or the Moonlight Project, and to organizations that are capable of receiving instructions from these experienced institutes or firms, for the reasons described above.

[8] The selection of collaborative R & D subjects and feasibility studies, and the education and training of staff starts in the 1993 financial year.

[9] When the necessary number of collaborative R & D participants cannot be found in the partner country, personnel from other countries who have sufficient technical background for the R & D work are welcomed, provided they are neighbors of the partner country.

At present, solar energy utilization technology (including a solar heating and photovoltaic generation hybrid system) is a candidate subject for collaborative R & D. In the future, other fields should also be considered.

4 The need for a systematic approach to energy conservation

In general, energy conservation has been largely concerned with the improvement of elemental technologies, like the quality of heat insulation or the efficiency of individual machines, such as exhaust heat boilers or heat pumps.

Although unquestionable achievements have been made in these domains, we have now reached a stage where further improvements call for a more systematic and functional approach to energy utilization.

Several measures can be taken in this direction. From an engineering perspective, this includes rationalizing the utilization of energy in its multiple forms and creating complex energy systems that encompass heat utilization and power recovery with maximized exergetic efficiency. The following systems deserve attention, as representative of effective energy conserving technologies:

- cascaded energy utilization systems
- co-generation systems
- unexploited energy utilization systems

The systematic combination of these three technologies enables the creation of large-scale, centralized electric power generation plants for a highly efficient and stable supply of heat, electricity and power. The fact that both cascaded energy utilization and co-generation systems cannot materialize without an efficient utilization of mid to low temperature waste heat further emphasizes the social importance of unexploited energy utilization systems, particularly from a long-term perspective.

4.1 The concept of cascaded energy utilization systems

The cascaded use of energy, ie multi-stage use of energy with consideration given to its quality, is extremely rational from a thermodynamic viewpoint. This concept is applicable to thermal energy (multi-stage use by temperature level) as well as heat-and-power (co-generation) and electricity (combined cycles and re-powering).

Future co-generation systems are expected to greatly benefit from an extensive use of fuel cells. Fuel cells are electro-chemical energy conversion devices for direct conversion of fuel chemical energy into electric energy, which feature high efficiency, low pollution, multi-fuel capability and ease of location and are expected to be applicable to a wide range of power generating systems, such as large-scale, centralized, suburban and on-site systems and other types. In Japan, fuel cell technology development has been proceeding with comprehensive R & D on fuel cell types such as phosphoric acid, molten carbonate, solid oxide and others.

For the phosphoric acid fuel cell, two different types of 1000 kW power generating systems and 200 kW on-site systems are being developed. The former has undergone a successfully rated output run in fiscal 1987 and the latter, expected to be put to practical use in the near future, is undergoing R & D on applications to isolated islands and commercial utilization.

The molten carbonate fuel cell is characterized by a higher efficiency in comparison to the phosphoric acid type and can be utilized with coal gas. Ten kilowatt class stacks have been developed for fiscal 1996, and the accomplishment of 1000 kw class power generating systems is expected for fiscal 1995.

The solid oxide fuel cell, which may be the most efficient in the future, is undergoing development on elemental technologies. The present status and development targets for fuel cell technology are summarized in Table 3. Also, the trends of power generation efficiency improvement of combined cycles are shown in Table 2 as a reference.

In order to apply the concept of cascaded heat utilization, efforts should be made to form the ideal energy flow among industrial processes classified by their temperature levels, including lower temperature levels below environmental conditions, such as LNG. In Japan, LNG is stored at the temperature of -160°C and is usually evaporated by sea water, but this energy should be recovered by industrial processes at its highest efficiency limit. With this point of view, the author suggests a representative example of ideal exergetic utilization with a cascade form as shown in Figure 6.

An effective utilization of this system requires the incorporation of co-generation, heat pump and energy storage technologies. Predicted rises in electricity demand coupled with difficulties in long-distance heat transport further enhance the merits of this system. Figure 6, needless to say, assumes the existence of a baseload supplied by a large-scale, centralized electric power plant. The current need for effectively utilizing the vast amounts of waste heat that result from electric power generation may make it necessary to scale down power plant sizes to a certain extent, and locate them in the vicinity of areas with high heat and electricity demand. Typically, co-generation systems are broadly divided into topping cycle and bottoming cycle, the former involving direct generation of electricity by fuel combustion with gas turbines or fuel cells, and the latter consisting of power recuperation from waste heat.

In the case of Japan, topping cycle co-generation has rapidly diffused into the industrial sector, while an expansion into the commercial/residential sector requires support by regulatory provisions and technological advances in areas such as heat/electricity ratio control and freedom from maintenance. Another important element regarding the energy utilization situation in some countries is the distinction made by the government between primary energy supplied for Electricity Generation and that for purposes Other than Electricity Generation. In Japan, 37% of the primary energy (nuclear, geothermal, fossil fuels and others) is utilized for electricity generation, and the remaining 63% is used for purposes other than electricity generation, as shown in Figure 3a. If energy utilization is to be rationalized by adopting co-generation and cascaded use technologies, however, this parallel distinction between energy for electricity and energy for other purposes must be thoroughly re-formulated.

The basic concept of cascaded energy use involves fully harnessing the heat generated by fuel combustion, from its initial 1800°C down to near environmental temperatures, with a thermal 'downflow' of heat analogous to the downward flow of water in a cascade. One obvious example of such utilization is the use of co-generation systems where a heat engine generates power and converts it into electricity, and the exhaust heat is utilized in industrial processes and/or in heating, cooling and hot water supply systems for the commercial/residential sector. Such systems, where both heat and power/electricity are generated and used, can only be fully efficient if electricity and heat are managed not in the conventional 'parallel' way, but in a 'serial' manner. As cascaded use and co-generation systems such as fuel cells become more widespread, the cases of simultaneous heat and electricity generation at the same location tend to increase.

While heat demand in industries and hotels, for example, has hitherto been met by using boilers, the advent of cascaded energy use will require the use of engines instead of heat-only generating boilers, which are no longer a thermodynamically sound option. A more widespread use of engines as heat sources will mean more electricity generated as heat is produced, which will require careful re-formulation of the future energy supply structure to accommodate these much needed changes.

As has been mentioned earlier, a highly important task from now on entails determining how unexploited energy, including low to mid temperature waste heats can effectively be utilized within such systems. Significant developments in energy conservation by waste heat recovery can be achieved if near environmental temperature waste heats can be effectively utilized as heat sources for heat pumps. While heat pump technology has an enormous potential in terms of CO_2 reduction, developments in bottoming cycle co-generation are also highly important for effectively utilizing large amounts of industrial waste heat. A systematic combination of heat pump and co-generation technologies is thus highly desirable from the viewpoint of establishing a free interchange of energy between different economic and geographic sectors.

The author firmly believes that we are now at a time when energy utilization must not confine itself to one specific sector, but rather be functional as a system, in an organic combination where mutual profits are ensured to a multitude of corporations. Positive action in promoting the development of new integrated industrial/commercial community projects is essential in the establishment of new environment-friendly communities with minimized exploitation of energy resources.

Furthermore, if future needs for CO_2 fixation and recycling are considered, as shown in Figures 7 and 8, the development of poly-generation systems may contribute not only to energy conservation but also to environmental preservation. Figures 3b to 3f outline possible cascaded systems for utilizing energy in a serial flow manner.

4.2 Technological breakthroughs for unexploited thermal energy use

Among the energy systems shown in Figure 6, one aspect which deserves particular attention is the rational utilization of near environmental thermal energy to meet the heat demand of the commercial/residential sector as shown in Figure 3f.

In the case of Japan, for example, the growth in heat demand for HVAC and hot water supply systems in this sector has been exceeding twice the national growth rate of energy consumption. It must be noted, however, that most of this demand can be met by heat sources of relatively low temperature, below 100°C. Hence, rather than using high grade energies as heat sources it can be very appropriate, also from a thermodynamic viewpoint, to construct systems capable of utilizing hitherto overlooked and unexploited low grade energy sources.

If one looks at urban areas with high heat demand, one will find that this demand can be met and even exceeded by utilizable yet unexploited nearby energy resources. These include not only waste heat from electric power generation and garbage incineration plants, but also river and sea water—which have temperatures differing from the atmospheric throughout the year—as well as sewage, for example.

High temperature heat supply, which utilizes steam or other latent heat media and has a broad application range, is generally regarded as the most desirable heat transport system. If, however, heat demand in the non-industrial sector is emphasized, from a different perspective, it becomes possible to meet most of the demand with low- to mid-temperature waste heat. Moreover, by making extensive use of absorption and super heat pumps, it is possible to broaden the range of unexploited heat utilization and effectively use low temperature heat sources and heat sinks at near environmental temperature, such as river and sea water.

In an effort to construct district heat supply systems utilizing unexploited energy sources, the Japanese government (through MITI) has been promoting R & D on large-scale, high efficiency machines, heat transport technologies, large-scale heat storage and the planning/operation of optimized systems combining these technologies. An expansion of district heat supply systems through dramatic increases in the efficiency of unexploited energy utilization systems is expected to assist in reducing and optimizing energy consumption, and hence in peak-shaving the electric power demand, particularly by the residential/commercial sector. An example of a development project incorporating these concepts is shown in Figure 5. The project is scheduled for implementation between 1991 and 1997. Design and manufacture will be completed by 1994, and verification studies will begin in 1995.

The planning and establishment of such unexploited energy networks as part of the social infrastructure require the solution of several problems, the main technical ones being summarized on Table 4. These technological improvements must not be restricted to the establishment of criteria for standardization and the development of elemental technologies. They must also involve a better understanding of overall system performance improvement by means of empirical experimentation. In particular, substantial contributions to the diffusion of unexploited heat utilization systems can derive from standardization and the resulting hardware cost reduction, provided that the temperature levels of collected and discarded heat are adequately matched upon the planning and establishment of such broad networks.

5 Concluding Remarks: the Establishment of an International Energy Infrastructure

In the preceding section, clean energy technologies and integrated energy utilization systems were presented and discussed, with emphasis on cascaded energy utilization concepts and the need for a systematic approach to technological development in this field.

Energy infrastructure provision is mostly important and effective with regard to the construction of such rationalized energy systems. The present can be said to be the time for preparing a high quality energy infrastructure to bequeath as a social capital to future generations.

In terms of electric energy, meeting the needs of future increases in demand will only be possible with the provision of ultra-high voltage electric cable networks, which will undoubtedly benefit from the development of ambient temperature superconducting materials. Furthermore, in urban areas, where heat demand is highly concentrated, there is an urgent need for the provision of a heat supply infrastructure capable of effectively utilizing unexploited energies such as river water and sewage, as described in Section 4.2.

While the provision of such energy infrastructure within each country can now be considered as a matter of course, the future will call for the development and planning of international gas pipelines and electric cables. For example, after the European unification in 1992, a rationalized electricity and gas supply system may materialize within the European Community.

In response to this international state of affairs, the establishment of an international electricity and gas network can be conceived, including technology transfer from industrialized to developing countries. From a long-term perspective, such a structure is effective in terms of ensuring energy security in countries with limited energy resources. It would also contribute to the technological progress of developing countries, thus playing a significant role in the technology transfer scenario.

A further and most important impact can also be achieved in global environmental protection, from a substantial reduction in the emissions of CO_2 and environmental pollutants from the developing countries.

In conclusion, the author would like to stress that the time has come for expanding our international vision to take in the development of energy infrastructure.

Table 1
Atmospheric concentration and life span of greenhouse gases

ppbv: parts per billion in volume

MICRO-CONSTITUENT GASES	CONCENTRATION BEFORE INDUSTRIAL REVOLUTION	CONCENTRATION IN 1980	LIFE SPAN IN ATMOSPHERE
CO_2 (CARBON DIOXIDE)	275×10^3 ppbv	339×10^3 ppbv	7~14 years
CH_4 (METHANE)	600 ppbv	1550 ppbv	10 years
N_2O (NITROUS OXIDE)	280 ppbv	301 ppbv	170 years
CCl_3F (CFC R-11)	———	170×10^{-3} ppbv	65 years
CCl_2F_2 (CFC R-12)	———	280×10^{-3} ppbv	120 years
$CFCl_2CF_2Cl$ (CFC R-113)	———	30×10^{-3} ppbv	90 years
CH_3CCl_3 (METHYL CHLOROFORM)	———	140×10^{-3} ppbv	6~7 years
$CHClF_2$ (HCFC R-22)	———	60×10^{-3} ppbv	20 years

Table 2
Power generation efficiency improvement of combined cycle systems: technological trends

Technology	Development	Present status	Targets(yr.2000)
High efficiency gas turbine	Efficiency improvement by high temperature combustion	Electric power generation efficiency:43%	Electric power generation efficiency above 47%
High temperature. high efficiency gas turbine for power plants	Development of new heat-resistant materials	(Combined cycle) (¥150000/kw) 1100℃ operation	(Combined cycle) (¥180000/kw) 1300℃ commercial machin 300kw commercial machine (1996)
Coal gasification combined cycle power generation	High performance gas turbine & gasification furnace	Pilot plant under construction (Coal gasification furnace,200t/day)	Electric power generation efficiency :46% (Conventional fine coal combustion:37%)

Table 3
Power generation efficiency improvement by fuel cells: technological trends

Phosphoric Acid Fuel Cell		
Items	Present status	Targets for yr. 2000
High efficiency technology High temp. & press. type Atomospheric temp. & press. type	Power generation efficiency 38% 36%	Power generation efficiency 40~42% 37~38%
Large-scale technology High temp. & press. type Atomospheric temp. & press. type	1000kw Power generation system 200kw	5000~10000kw class power generation system 1000kw class
Low cost	1,000,000~2,500,000 yen/kw	200,000~250,000 yen/kw Level
New Type Fuel Cell		
Items	Present status	Targets for yr. 2000
Molten carbonate fuel cell	100kw Development under way Elemental technology development under way	Power generation efficiency above 45% 1000kw Class development (1996)
Solid oxide fuel cell	Feasibility study Elemental technology, development under way (1.2kw)	Power generation efficiency above 505 Bottoming cycle overall power generation efficiency above 80% (Combined with high temperature waste heat-drive turbine)

Table 4
Technological developments required for unexploited energy use

main technologies	elemental technologies	development needs	comments
waste-heat recovery	◊ heat exchangers for small ΔT - *plate type heat exchangers* - *heat transfer enhancement*	◊ high efficiency plate-fin heat exchangers ◊ stainless steel,ceramics,glass heat exchangers; anti-corrosive coatings	◊ cost reductions by mass-production
	◊ corrosion-resistance	◊ maintenance-free, cleaning	
heat storage	◊ optimum heat-storage systems for each temperature level (mainly sensible-heat storage) - *chilled-water cold storage (5–10°C)* - *unexploited heat storage (10–45°C)* - *hot-water heat storage (45–50°C)*	◊ large-scale heat storage tanks 40–90% heat recovery efficiency or better	◊ underground water, rocks & other natural media can be used
	◊ latent-heat storage - *ice thermal storage (0°C)* - *PCM thermal storage (-10–80°C)*	◊ reduced resistance to heat transfer - *direct-contact type (slurry)* - *supercooled type* - *clathrate type (contact temperature 7°C)*	◊ maximum heat-storage density ◊ constant-temperature heat storage
	◊ chemical heat storage - *high-temperature heat storage (80–200 °C output temperature)* - *cold-storage (output temp <10°C)*	◊ selection of media applicable over a broad temperature range - *ammonia use* - *high-efficiency, large-scale reactors*	◊ in the developmental stage
heat pumps & chillers	◊ compression heat pumps & chillers - *compressor technology* - *new working fluids (non-azeotropic mixtures)*	◊ high efficiency & high temperature lift - *COP 3–5 or better* ◊ no-freon refrigerant use started	◊ compactness ◊ maintenance-free ◊ high reliability
	◊ absorption heat pumps & chillers - *advanced cycles* - *compression/absorption hybrid cycles* - *new working fluids*	◊ high efficiency & high temperature lift - *COP:>2–3 for heating; >0.6–2 for cooling* ◊ advanced cycle demonstration - *GAX, dual, triple, etc* - *low-temperature drive & ammonia use*	◊ can be directly driven by unexploited thermal energy ◊ most suited as no-freon type
	◊ adsorption heat pumps & chillers - *high-efficiency reactors* - *new adsorbents* - *new cycles*	◊ high efficiency - *cooling COP of 0.6–0.8 or better* - *heat transfer enhancement inside reactor* ◊ development of continuous cycles	◊ directly driven by heat sources of lower temperature than absorption chillers ◊ compactness & lightness problematic
piping works	◊ new piping construction methods - *shield tunnel* ◊ anti-corrosion pipe materials - *concrete tunnels, etc*	◊ lower pipe installation costs ◊ good anti-corrosion pipe materials with heat-storage capability	◊ cost in urban areas is 10^5–10^6 yen/m

Energ.Saikouchiku table

Figure 1
Systematic overview of environmental problems

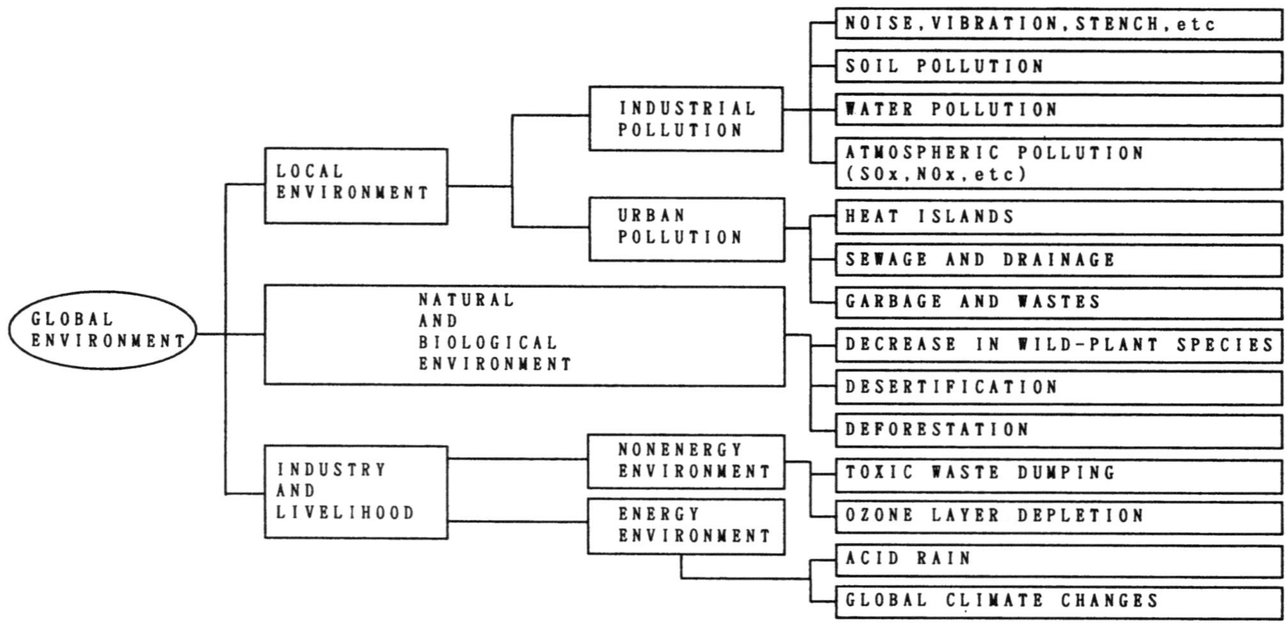

Figure 2
The beginning and the end of interglacial periods

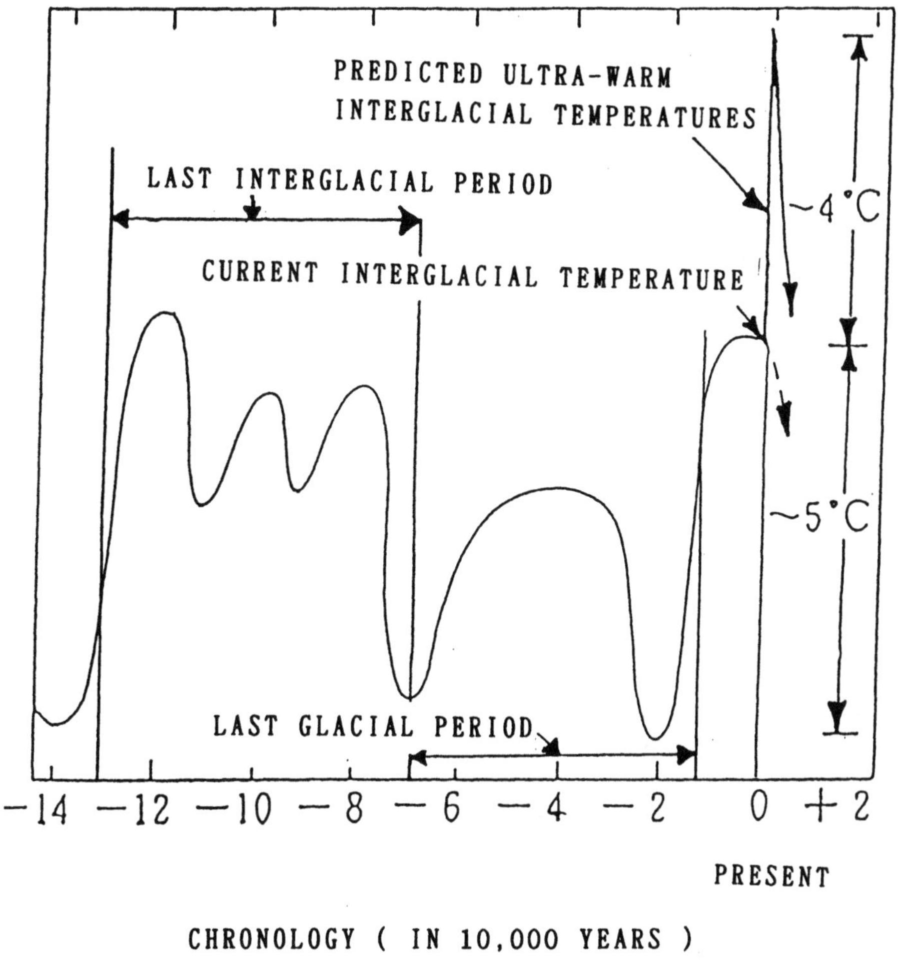

Fig. 3d
Facilities for electric power generation from wastes
Stimulate the use of heat from waste-incineration, which traditionally has not been harnessed as energy.

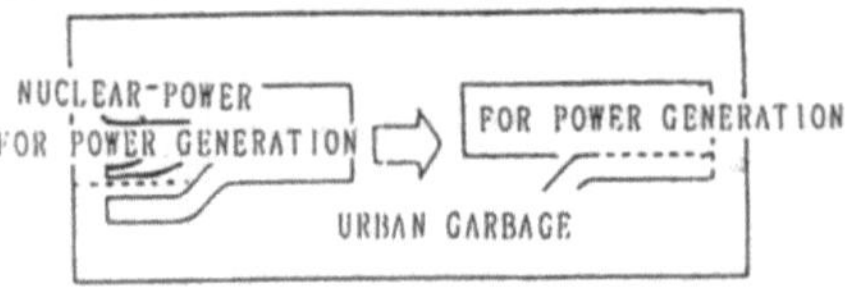

Fig. 3b
Large-scale district supply of co-generation heat
Expand the use of waste heat from power generation by locating co-generation systems in the vicinity of areas with large heat demand.

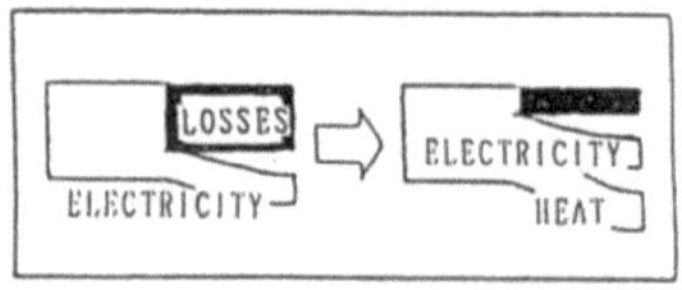

Fig. 3a Flowchart of energy supply and consumption in Japan (F.Y. 1986)

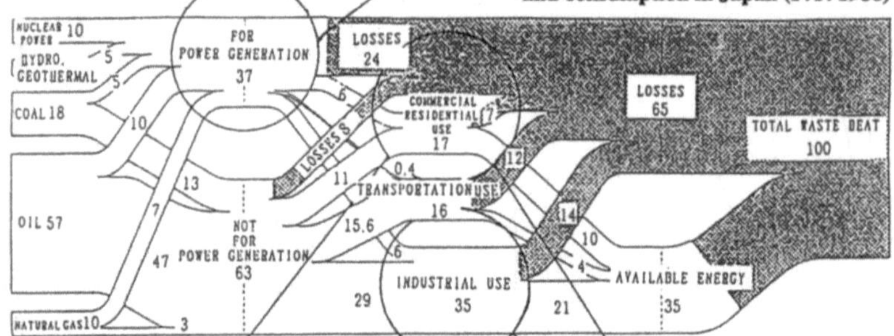

Fig. 3f
Facilities for district-level heat supply utilizing unexploited energy
Reduce the input of new energy by utilizing unexploited energies such as temperature differences in river and sea water, as well as urban waste heat from subways and other sources.

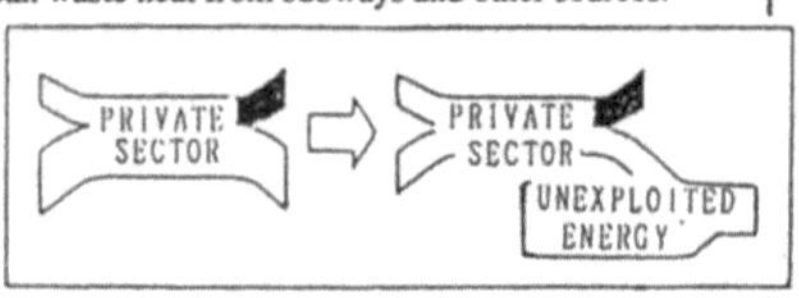

Fig. 3e
Facilities for supplying surplus energy from power plants and factories
Reduce the input of primary energy by effective utilization in the commercial-residential sector of waste heat not suitable for industrial use.

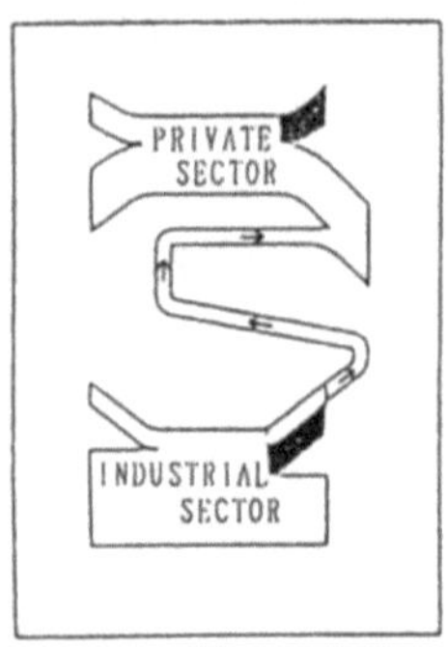

Fig. 3c
Industrial complexes with cascaded energy utilization
Reduce the total input of primary energy into industries by promoting the multi-stage utilization of heat.

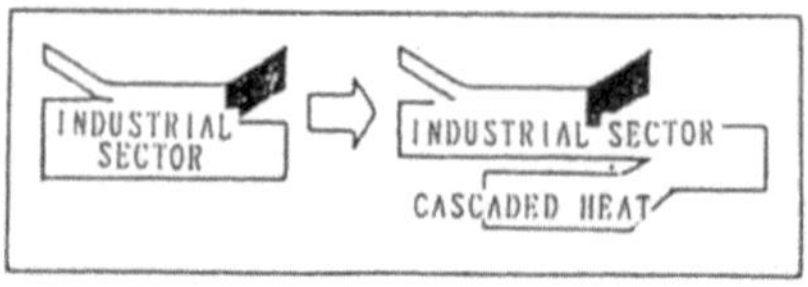

Figure 4
Example of an integrated energy system

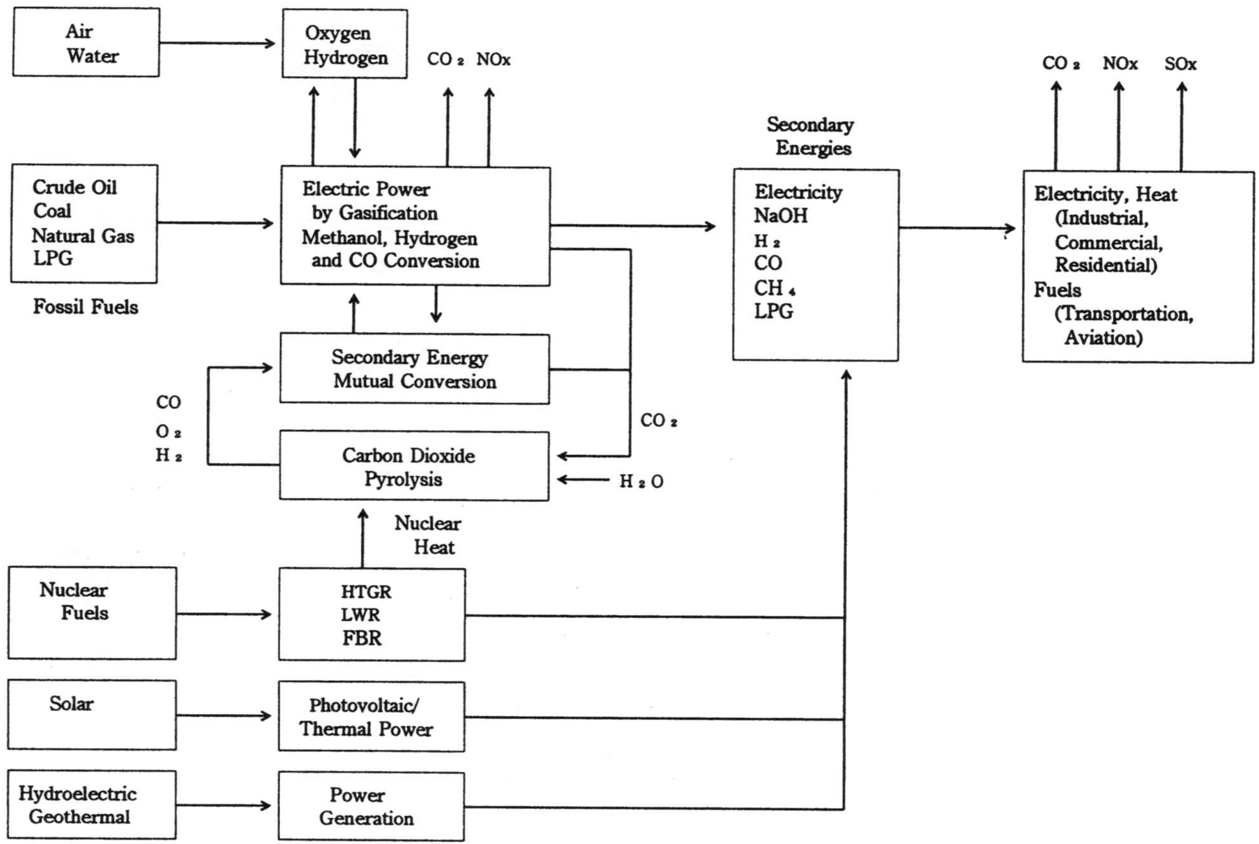

Figure 5
Load-levelling air-conditioning system using unexploited energy

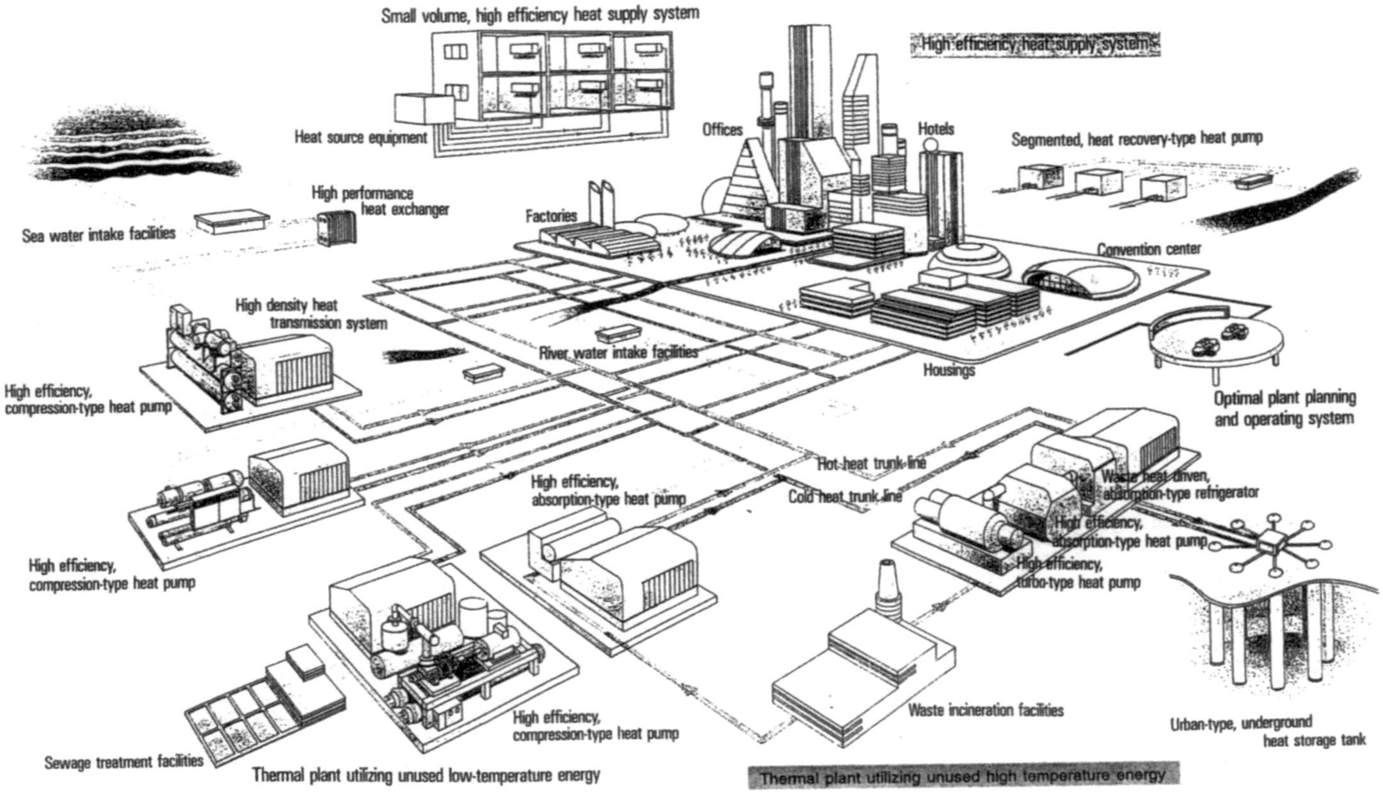

Figure 6
Cascaded energy utilization concept and systems integration

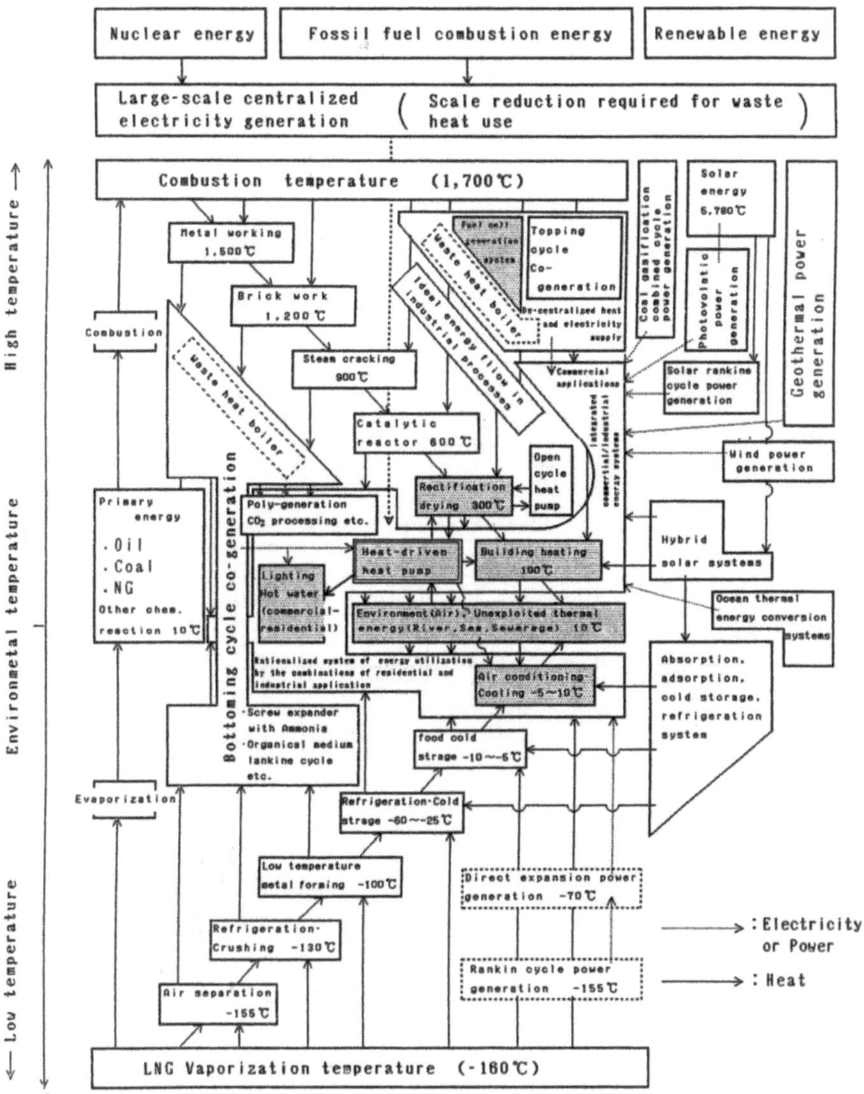

Figure 7

CO_2 fixation and re-utilization technology: photosynthetic microorganisms and algae

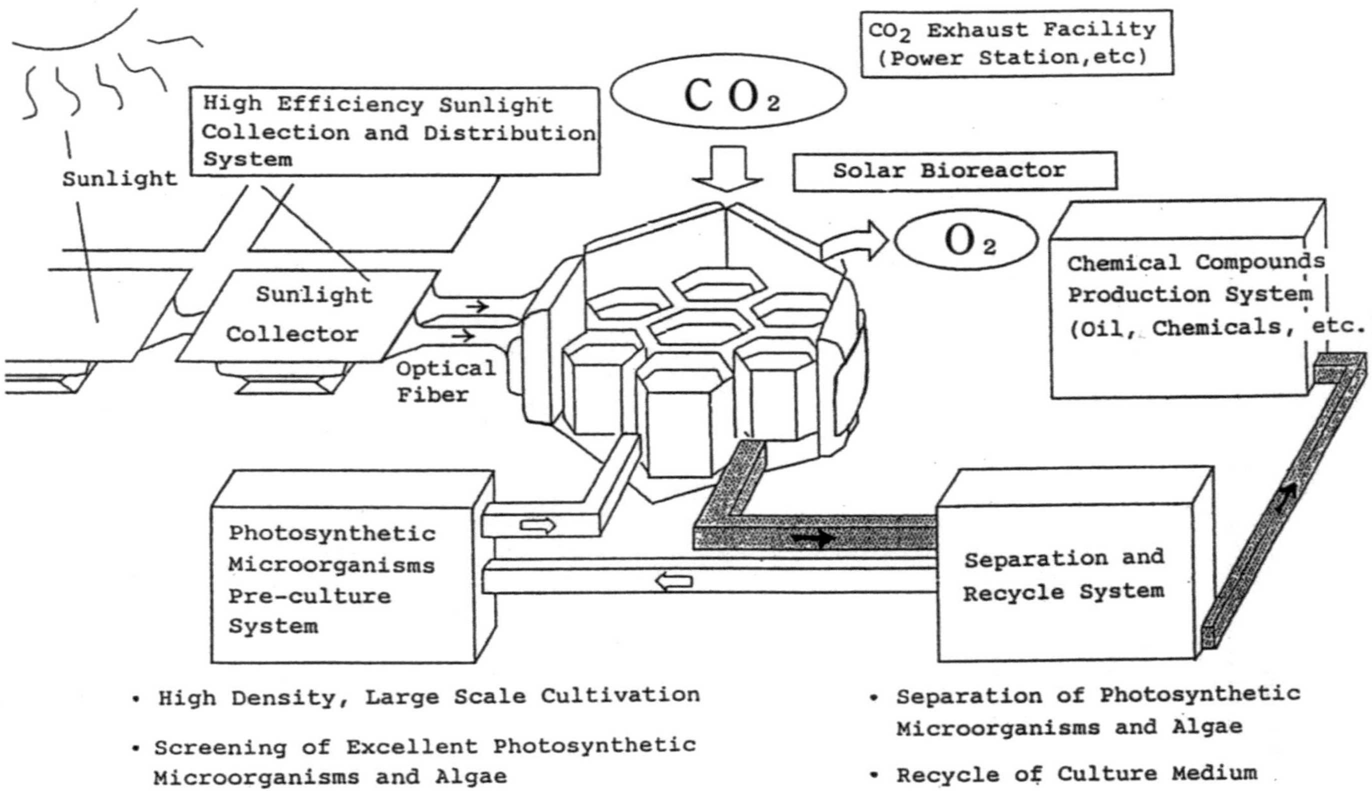

54 GLOBAL ENVIRONMENTAL SECURITY

Figure 8

CO_2 fixation and re-utilization technology: catalytic hydrogenation

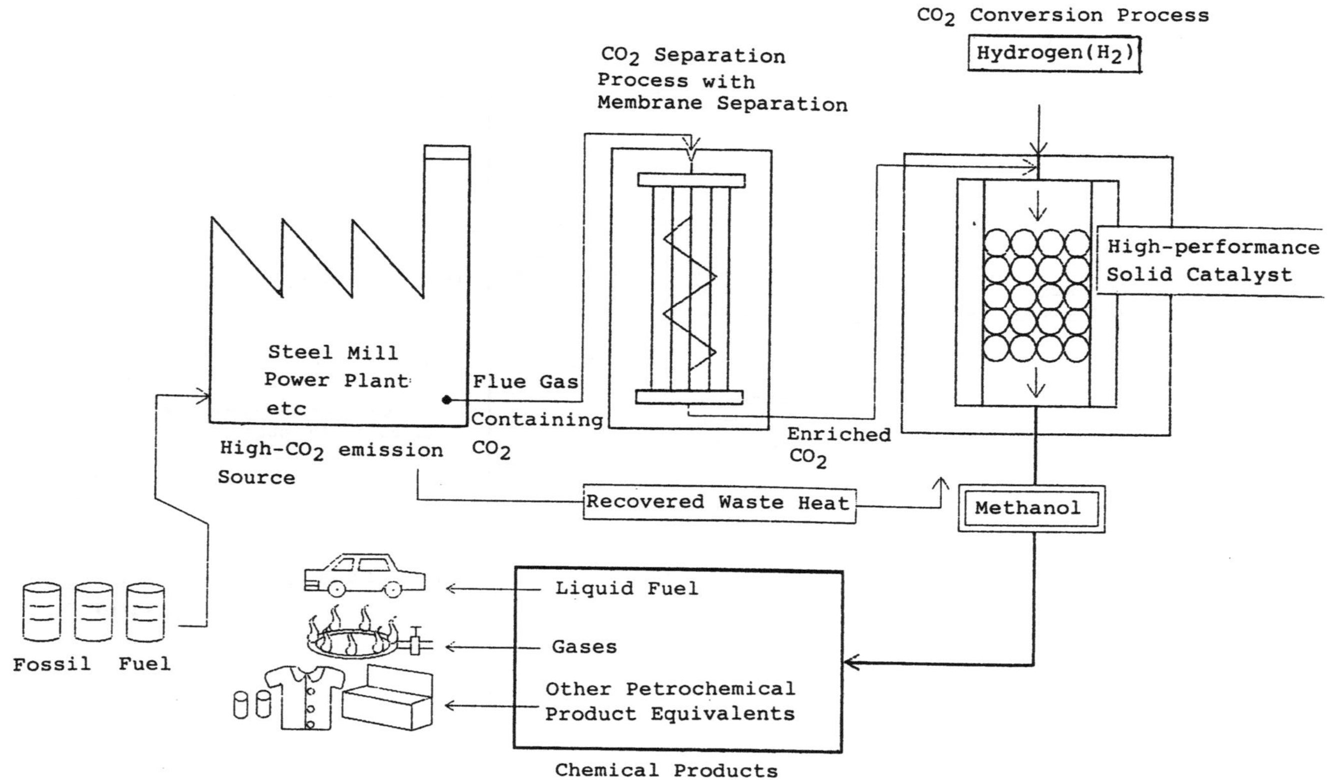

Figure 9
The New Sunshine Program system

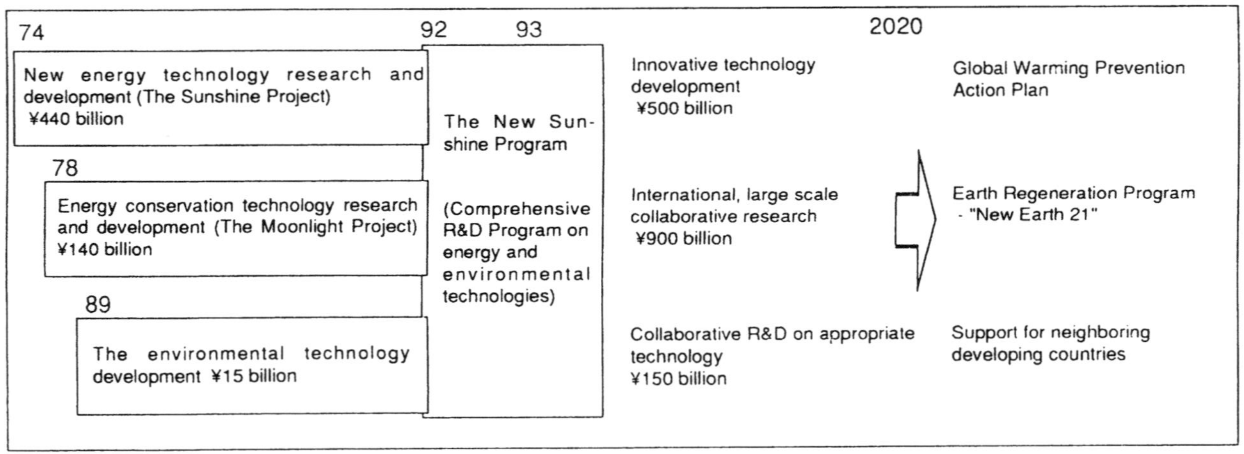

Figure 10

Contribution of the New Sunshine Program to the Earth Regeneration Program

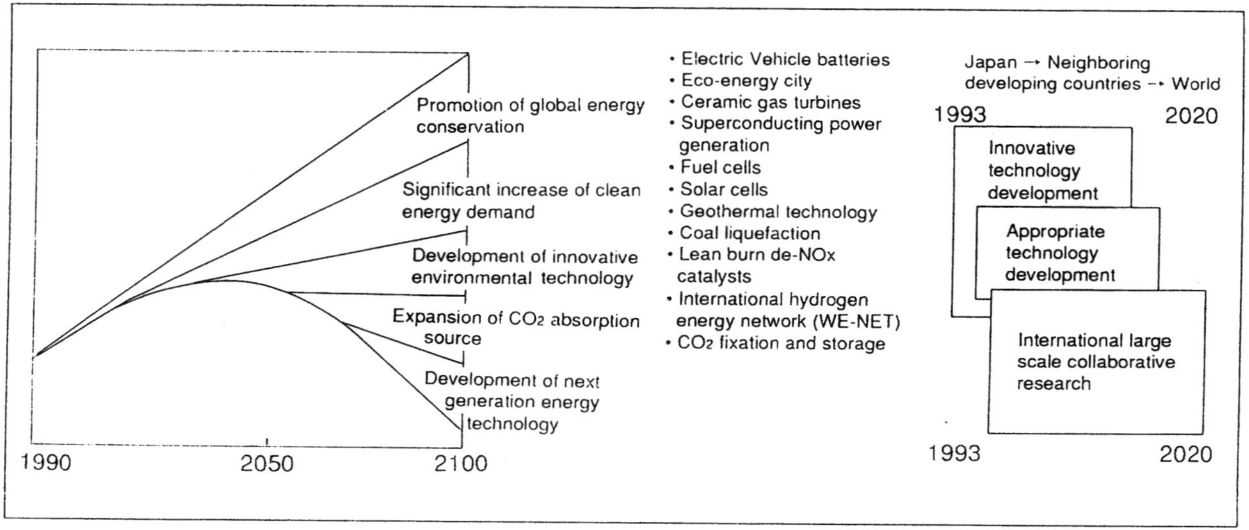

Figure 11
Contribution of the New Sunshine Program to easing energy demand/supply and CO_2 restrictions

Contribution rate to easing energy demand/supply

Beginning of 2000	The 2010s	Around 2030
Several %	Approx. 10%	Approx. 34%

Contribution rate to easing CO_2 restriction

Beginning of 2000	The 2010s	Around 2030
Several %	Approx. 16%	Approx. 50%

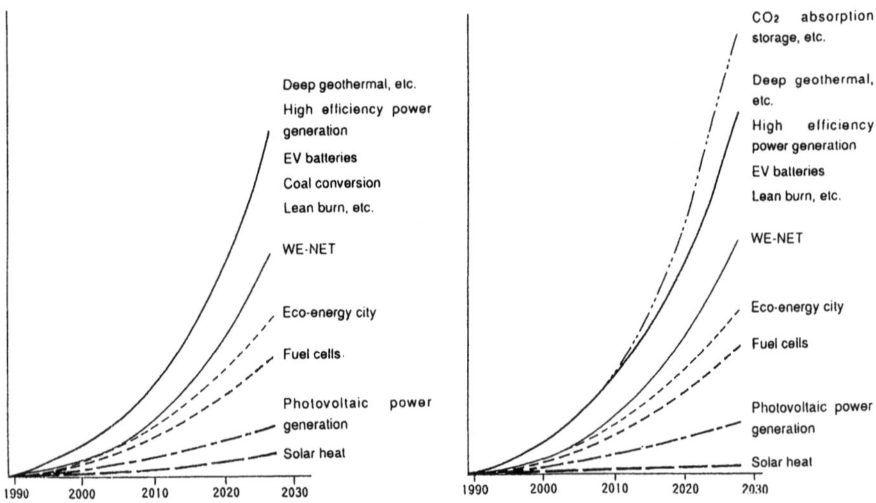

(Notes)
1. Energy: (Quantity of energy conserved + Quantity of new energy supply) / Total energy supply
2. CO_2: (Quantity of CO_2 emission reduced + CO_2 absorbed and stored) / Total CO_2 emitted
3. Technological potential of the maximum effort for implementation of the New Sunshine Program.

Figure 12

Expected accelerated promotion of research and development and resulting cost reductions

1) Solar cell

Improvement of conversion efficiency, mass-production, cost reduction technology

	At the end of FY'92		At the beginning of 2000
Total system	Approx. ¥450 /W	→	Approx. ¥220 /W
Cell	Approx. ¥330 /W*	→	Approx. ¥150 /W
Peripheral equipment	Approx. ¥120 /W	→	Approx. ¥70 /W
Power generation cost	Approx. ¥40 /kWh*	→	Approx. ¥20 /kWh

* As of 100 MW/Y*plant of production scale

2) Fuel cell

Cost reduction, compacting technology, validation operating study

	At the end of FY'92		At the beginning of 2000
Total system	Approx. ¥800 /W	→	Approx. ¥300 /W
Cell	Approx. ¥240 /W	→	Approx. ¥90 /W
Peripheral equipment	Approx. ¥560 /W	→	Approx. ¥210 /W
Power Generation Cost	Approx. ¥40 /kWh	→	Approx. ¥20 /kWh

Figure 13
Schematic of the broad area energy utilization network system

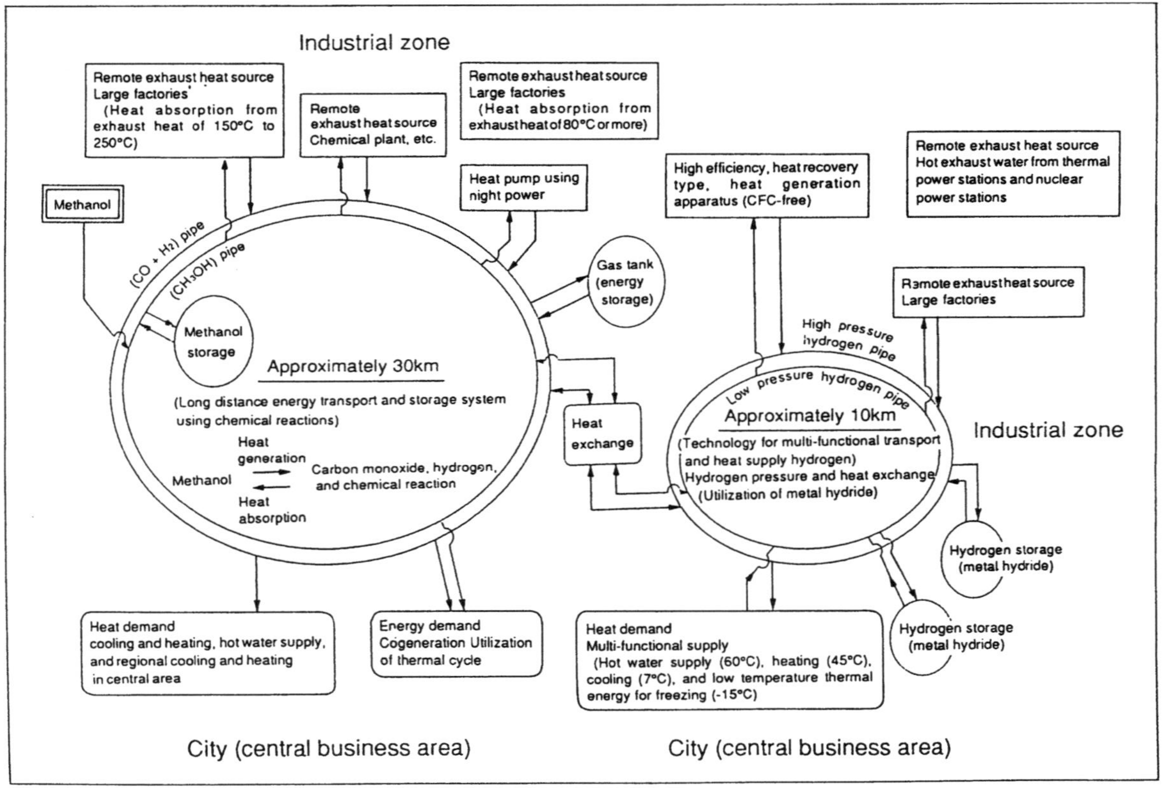

Figure 14
Concept of WE-NET

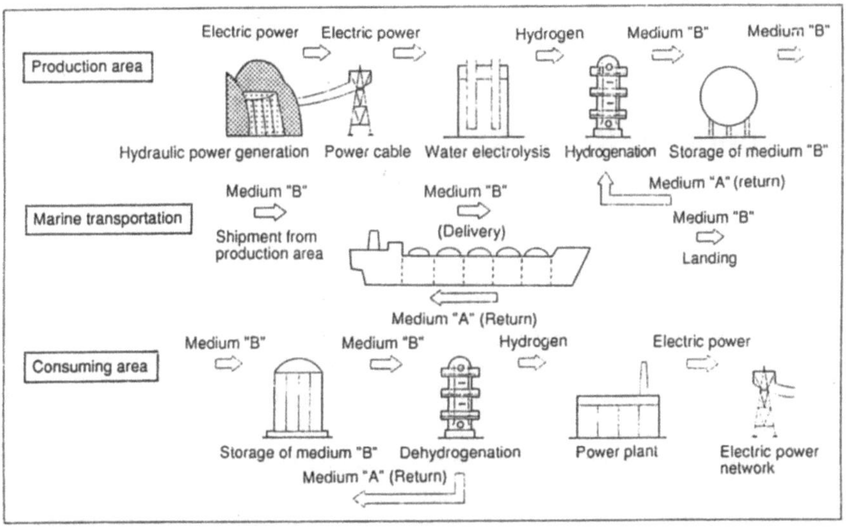

Figure 15
Concept of exhaust gas de-NOx catalyst technology for lean-burn engines

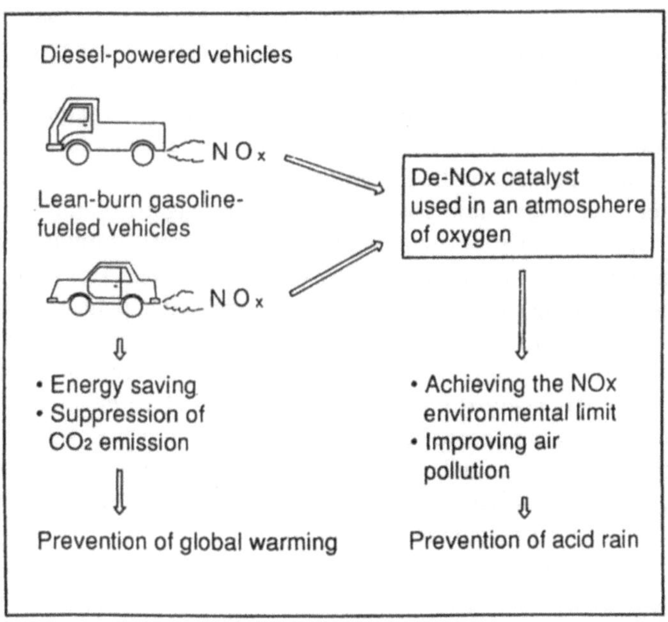

I.5 Appraisal Remarks

K. Takeuchi

The purpose of my talk is the appraisal of the present state of the global environment. Also, I would like to touch on the scope of our common efforts, cutting across divisions including North and South, toward sustainable development. Of course, I do not pretend to have any definite proposals or answers. Based on my own personal viewpoint, the main question is how to formulate problems and in what kind of framework we are going to look at them.

After the Rio Summit of 1992, there is bad news and good news. The bad news is the fact that after the Rio summit, public interest and attention has been a great deal in decline and diverted to other places and problems. But there is also good news. Sound and solid progress has been made toward better understanding of environmental issues. Also a number of concrete policies have been put into effect toward solving these issues.

Since 1992, maybe from a little earlier, economic recession has taken over public interest in many industrial countries, especially Japan. Global environmental issues have been pushed aside and some say that these are a kind of luxury that maybe we can indulge in when we have economic prosperity. I think this is very bad news. In addition, there have been social troubles and ethnic wars, for example in former Yugoslavia and some parts of Africa. Of course wars and social unrest are the worst thing that can happen to not only human welfare but also to the environment.

Another piece of bad news is that some are now eagerly promoting economic policies which stimulate economic growth irrespective of environment, with more consumption of expensive energy and other wastes. The good news that basic understanding has been promoted and some actual policies implemented has been recently discussed. I think that present governments seem to have a better understanding of the issue of environmental protection. Also, I can say there have been many things which have been clarified in the course of discussion and in the course of actual processes. There are two aspects here. One is the contradiction and the other is unity or harmony. The contradiction is to be seen in various issues of environmental problems and especially in the concept of so called sustainable development.

There are also contradictions found between different levels of environmental problems. We can distinguish different levels of environmental problems as local, regional, national and global—and also as short-term, mid-term and long-range. Some may start at a local or regional level and then move first to the national and then to the global level. Some problems may start out appearing to be short-term and become something that endures for centuries. Sometimes it happens that something good for local environment can be bad for global environment. A typical example is CFC. CFC has been considered to be very good to the environment because it doesn't do harm at the local level, because of the fact that it is chemically stable. However, it is seriously damaging to the world globally since it can destroy the ozone layer in the stratosphere, thereby causing warming. This may be beneficial to some parts of the world in the short term but is going to cause serious problems in the long term. This is one kind of contradiction. The second is of course the relationship between economic growth and environmental protection. For some time there has been talk of a trade-off between the two, but I believe the actual contradiction is quite misunderstood. I would like to talk more on this a little later.

Another type of contradiction occurs when international politics stands in the way of the international cooperation vital for the wellbeing of the global environment. Here I believe fundamental unity for the sake of global environmental wellbeing is basic, because all human civilization is, in the long run, interwoven and cannot be separated. In addition, natural sustainability and social sustainability are directly connected. Dr Salim explained very clearly about the conditions of social sustainability which not only include the necessity of providing people with economic welfare but also with the means of achieving social justice, equality and so on. Fundamental unity must be emphasized more.

I would like to propose a basic motto here, that nothing that is bad for human welfare can be good for the environment, and nothing that is bad for the environment can be good for human welfare. For example, economic depression cannot be good for the environment either. Some who talk about the trade-off between economic growth and environment conservation as being unavoidable say that economic depression is good for natural environment because decreasing economic activities and energy consumption result in a decrease in those activities that destroy and pollute the environment. But anything which causes misery for people cannot be good in the long run for environment preservation. Poverty and economic depression prevent us from building up the capability to cope with problems such as those affecting the environment. Neither can war and social disturbances be good from any viewpoint.

Destruction of the environment cannot bring any economic benefit either, except for very small groups for a very short time. Dr Brown stated, in quoting many examples, what is wrong and what has been badly done. We have to find some harmonious ways to improve human welfare and strengthen the conservation of the natural environment.

In order to harmonize, we have to recognize three or four basic factors. The first is that we have to clearly recognize the human needs of people. Human need can, and should be, distinguished from economic or political demands such as the growth of GNP. Recently, in Japan, a book called 'Philosophy of Being Clean and Poor' sold seven hundred thousand copies. It indicates that the people in Japan have understood that simply being rich in monetary terms does not bring happiness. I don't want to preach asceticism but it is important to distinguish between economic demand and human need.

Secondly, we should be very realistic about the time span of the problem. We have to distinguish between long-term and short-term problems and be clear about what we can do in the long range and in the very short range. Some people say that since the global environment problem has a very long-range perspective we have to do long-range planning. But it is not realistic to make predictions or plans beyond maybe thirty years from now. It is possible, though, to plan for a realistic period and at the same time to look ahead maybe two hundred years into the future.

Of course, it is necessary to somehow stabilize the conditions on the surface. But stabilization often means status quo and we also have to have dynamic change. The key factors for the solution of global environmental issues are therefore:

First of all, to enhance technological capabilities and to develop potential technologies.

Secondly, and of no less importance, is the need for capital investment and its proper allocation. We need a large amount of capital investment in the related technological developments. Also important is that human resources must be effectively utilized. Human resources here means not labor force but human inventiveness and human aspirations. For this end, peace—international, domestic and social peace—is the key factor—cooperation among nations, regions, sectors of industries, and different sections.

Thirdly, we need a consistency in our activities that provides harmony but not uniformity. Perhaps we need a variety of actions suitable for different situations in order to achieve the same goals. Certainly we have to use different methods and policies in different situations. Also necessary is the coordination of activities by different peoples, agencies, and nations. These are the kind of general principles we have to apply in dealing with the environmental problems.

But as an economist, I would like to add a bit on economic policy. The world economy, I believe, is in transition from that of the Cold War to a kind of peace world. Today many nations are in the midst of an adjustment process in which not only the former socialist states but also the industrial economies face many new difficulties. In that process, I believe, environment protection should have a high priority.

Nonetheless, if effective economic policy measures are not taken now then some mismanagements could cause common problems such as in the 1930s. Policy adjustment and economic restructuring, therefore, are basic in this post Cold War period in which policies are to be formulated.

Secondly, we have to promote a kind of system for international cooperation by which not just peace-keeping but also environmental conservation can simultaneously be secured. National sovereignty must be superseded by international cooperation and aid programs directly and indirectly connected with environment conservation.

Finally, we need to construct a worldwide structure which would coordinate energy programs with other factors including communication, transportation, education and so on. Indeed, ironically, one of the merits of imperialism is that imperial powers have actually constructed a worldwide structure in which many things mentioned in the above could be combined, although only for imperial purposes.

In the post imperial era, I think, what is most important is to increase public awareness and consciousness with respect to the problems we are all faced with.

Thank you very much.

I.6 Panel Discussions

Originally we thought about the possibility of asking each participant to extend their views, but I think I understand that some listeners are somewhat frustrated, so it is perhaps legitimate for us to follow some suggestions that the floor should be open now. Secondly, this is not the sort of session to lead to concrete proposals. We have another day to debate. Now I would like to open the floor.

Dr Kato

I used to work for the Environmental Agency for Japan. All the presentations are very interesting, but let me ask Mr Lester Brown about so called broad based energy tax. In the United States, the Bill was said to be passed by Congress, but unfortunately it was reduced to a great extent. Do you have any comments on that?

Dr Nakamura

I have one specific question to Dr Salim about this regional cooperation for environment regulation harmonization. You seem to indicate here that regional cooperation may lead to discrepancies between regions. I wonder if you can elaborate on this, and on the desired direction concerning specific regional cooperation versus global environment cooperation?

Dr Cho (Korea)

I have a question to Mr Lester Brown and Dr Salim. Both of you seem to emphasize the importance of environmental stability. Then, what would be the role of governments, for example in the introduction of environmental tax and in subsidizing other activities?

Another area which requires government leadership is technology transfer, which I believe can be regarded as a contribution to the common interest.

Dr Salim

In my opinion, developed countries are very negative and defensive regarding the transfer of environmental technology to developing countries.

Dr Levine

Dr Brown proposed an environmental tax to take the place of other taxes, and presumably part of that tax might be recycled into environmentally useful activities. My question is whether any country has in fact tried that and if not, what country do you think might be most likely to try that as a pilot program?

Dr Kashiwagi showed on a chart a huge number of energy technologies for which research is happening. I would like your judgment about the two that are most likely to have pay-off in the next 10 or 15 years. Dr Salim gave a wonderful talk stating some of the most important principles involved in understanding sustainable development. I wonder, however, if you could see just two or three specific actions that need to be taken to make those principles start happening?

Chairman

Thank you very much. Are there any further questions? No questions. Okay. Now it is time for each participant to respond very briefly so that Dr Sprenger may raise more extensive comments later. Who will start?

Dr Sprenger

I feel it is very popular these days to talk about revolutions and we hear a lot about efficiency revolution and political revolution and now one key word seems to be environmental revolution. I have to admit I have been somewhat disappointed by the type of revolution you were discussing or proposing and especially disappointed about the means to implement such a revolution. In Mr Brown's presentation I got the feeling that your approach is very much supply-side oriented and technology oriented and I have my doubts about the way we can handle the problem by introducing new technologies.

I got the feeling with some of the examples you have given in your paper that we are just moving the problem from one environmental medium to another. You are talking about cars using electricity. I don't see how you can avoid the pollution problem by just shifting the problem from oil and gas to let's say, some coal-fired or other power plant that might generate electricity for cars. Then you have indicated some technologies which might be useful for some regions of this planet but which are not a solution for let's say, Europe and the more industrialized countries where climate makes it impossible to use, for example, solar energy.

You have mentioned the major discussion going on about relying on economic incentives but at the moment I don't see anyone among the industrial countries that might rely on taxes. I don't see any way to develop, from the economist point of view, a reliable, feasible and intellectually clean tax that could be used for thousands of products that are not environmentally benign. So, looking at the promises of the Clinton and Gore administration and the promises of the EC CO_2 Energy Tax, it sounds like the environmental version of the famous French theatre play 'Waiting for Godot' and this is waiting for an energy tax.

I also feel that, even though you mentioned at the very beginning of your paper that such revolution might need major shifts in human behavior and dramatic changes in values and lifestyles, this was missing in the presentation and the discussions. We need to look at the demand-side effects in the developing countries, and some examples have been given here, especially regarding China. I guess one of the major focuses of our activities, if we talk about global advantages, needs to be not only discussion of smoothing the environmental situation by thinking about technologies and the supply side but also thinking more dramatically about the demand side.

Chairman
Thank you very much indeed.

Dr Brown
A question on the question. It is not clear to me what you are referring to on the demand side. I did talk about energy efficiency, for example, as being one of the keys to an environmentally sustainable economy. Are you thinking about voluntary simplicity, lifestyle, adjustments or . . . ?

Dr Sprenger
I guess if you increase energy efficiency—let's say, the average consumption of gas for cars within the increase of cars—you might just end up compensating and have a lot of trade-offs. Looking at many of the industrialized countries which have had quite important energy efficiency gains, most of these gains have been compensated for by an increase in demand. If we look at the development in some countries—especially in Germany—we see what's happening in Eastern Germany, and the same will apply to the CE countries. I feel that all those energy efficiency gains would not be enough to compensate for the increases in demand which have a concrete effect on the pollution and environment.

Chairman
Now I will ask the panellists to respond as briefly as possible. There are many questions. Dr Brown, you can go first.

Dr Brown
In response to Mr Kato's questions about the Energy Tax and the Clinton administration, I think you are not the only one disappointed by the shortfall on achieving Energy Tax. We will have a four cent per gallon gasoline tax but this is not necessarily the last word. I did talk with the Vice-President a few minutes just before leaving on Friday night and there are many more things in the pipeline.

You may know that there is another, in some ways dramatic, tax proposal for financing the health care plan and that is a cigarette tax of $1 a pack and I think the chances of this tax being adopted are better than for the gasoline tax. If it is successful it will establish a precedent for the adoption of very stiff taxes and may break the mental barrier that now seems to exist in Washington. Indeed, tax has been made such a political issue of the last dozen or so years that politicians are very much afraid of it. If we have a breakthrough on the cigarette tax to help fund the expanded health care that could change the context within which we look at environmental taxes.

There are a number of other things that the administration is working on. They are using government procurement, for example, to set new standards for paper procurement, recycled paper with substantial post-consumer waste content, and new efficiency standards for computers in the U.S. There will be a move toward the use of natural gas powered vehicles to help lower air pollution and there is also a major green light program under way to get corporations and other organizations to replace traditional incandescent light bulbs with compact fluorescent bulbs.

On the environmental tax question raised by Professor Cho and Mark Levine, most people agree that it is one of those ideas that just makes sense. Now there are somewhere between 20 and 30 governments that have environmental taxes of one kind or another. A few in Europe have some carbon taxes—Finland, Norway and perhaps the Netherlands.

Chapter Two

Technology for Sustainable Development

II.1 Overview

This chapter contains four presentations on technologies for sustainable development. The first by M. Levine, having examined a number of old and new technologies for increasing energy efficiency, discusses the need for international collaborations, especially between North and South. Pointing out China as becoming one of the major concerns for global environment due to its rapid economic development, he argues that there is a huge potential for improving its relatively poor performance. For this, he demonstrates the importance of policy orientation toward energy efficiency in China on one hand, and on the other a program of international cooperation through technology transfer, capital investments and above all the exchange of ideas on sustainable development. The last analysis, touching on the possibility of achieving economic development as well as environmental protection, is the main point of interest.

Both the second by K. Yamada and the third by H. Ishitani provide the scope of possibilities which new technologies can offer to the ostensibly contradicting demands today, namely the ever increasing need for energy supply on the one hand and the more severe measures of environmental protection on the other. Yamada demonstrates the possibility of large scale utilization of photovoltaic systems and argues that this increasingly popular technology does have a tremendous future in terms of providing a kind energy source to the globe. Alternatively, Ishitani examines the possibility of technology for CO_2 disposal into the deep ocean. This paper provoked a number of counter-arguments, especially from those from the South. However, as he mentions in his paper, this attempt is important in developing technology for safer recycling of the end products, in this case CO_2—the technology for which remains still at an embryonic stage.

The last presentation is by R. Huang who touches on climate and environmental change in East Asia, or more correctly, China. As seen above, China is simultaneously experiencing rapid economic development and degradation of its eco-system. Having examined the widespread flood and drought patterns in China's recent past, he argues that a clear correlation between economic development and environmental problems can be seen. Because China's is a developing economy, what is needed is both international cooperation, to set an upper limit to global warming, and also concrete actions, within and outside China, for the resolution of such dilemmas.

II.2 The Need for New International Efforts to Increase Energy Efficiency: North/South Cooperation

M. D. Levine

Energy and Environment Division, Lawrence Berkeley Laboratory, Berkeley, CA 94720, USA

Abstract

Global problems of energy, environment and economics are becoming increasingly intractable. This is especially the case for developing countries, which simultaneously face problems of inadequate energy services, deteriorating environment and inadequate capital for essential investments. It is important to emphasize that energy efficiency is one of the very few approaches that can address these three issues in a significant way.

A study of historical trends indicates the important role developing countries will play in future energy growth and thr growth of carbon emissions. Some examples of both energy growth and energy efficiency opportunities from China illustrate many of the energy and environmental dilemmas facing the developing world.

Industrialized nations can increase energy efficiency and they should do so, but it is much more difficult for developing countries to make large improvements in energy efficiency without considerable support. It is in the self-interest of industrialized countries to provide such support for at least three major reasons:

- this is a highly cost-effective strategy for restraining the growth of CO_2 emissions;
- this is an essential strategy for reducing the growth of world oil consumption;
- it should provide markets for energy efficiency technology in the developing world.

It is thus essential to assess new ways in which the industrialized nations can provide significant support for energy efficiency initiatives in the developing world.

1 Introduction

Economic and social development will not occur in the world's lower income countries unless these countries are able to get substantially more energy services than they receive at present. Energy services enable improvements in labor productivity, added mobility and more comfort and convenience—contributing to the development process itself and to an enjoyment of the fruits of this process. Meeting the needs of the developing countries for these services over the next several decades, however, is likely to test the limits of the world's economic, political and environmental systems—and sustained development is in jeopardy as a result.

There are two ways in which the nations and peoples of the world can obtain greater energy services. We can either increase energy supply, using traditional or novel technologies and energy resources, or we can increase the efficiency with which energy is produced, delivered and used. A doubling of the efficiency of energy generation, transmission and use produces the same energy services as a doubling of energy supply.

The increasing need for energy services poses one of the great technological and policy challenges of the next several generations: increasing energy services for development in Asia, Africa, Latin America and the Caribbean, the Middle East and Eastern Europe—in many cases by multiples of four or more—during a time when conventional energy paths are no longer adequate, because they are too expensive and too damaging to the global environment. In essence, as a global society we need to find ways to deliver more services while at the same time shift from the delivery systems of the past to the systems of the future (Wilbanks, 1988).

The features of these systems of the future are still fuzzy at this point, as we learn more about the potentials of renewable energy technologies and other options, but one feature is clear. In industrialized and developing countries alike, the energy production and delivery systems must be far more efficient than they are now. Energy efficiency improvements will be a key, because in a great many cases they will deliver additional services more cheaply than supply additions, both economically and environmentally (eg Helsinki Symposium, 1990).

Moreover, we know that many of the potentials for efficiency improvement are technologically feasible and economically attractive right now, and further potentials can be made attractive with a reasonable amount of additional policy emphasis (Katzman et al., 1990). While the rest of the vision of our energy future comes into focus, this is one place where we can proceed with confidence: reducing capital requirements for expanding services, reducing environmental impacts associated with those services and encouraging more attention to the efficient use of resources as an integral part of any strategy for sustainable development.

This paper briefly explores efficiency improvement as an energy strategy for developing countries. First, it describes the evolving energy patterns in these countries and outlines the role that energy efficiency can play in the near and mid-range future. Next, it summarizes the main impacts of these patterns and the issues that must be confronted in improving energy efficiency in many countries, together with a growing record of success in this regard. Finally, it suggests the need for bold, new cooperation between North and South if a global effort is to be undertaken to accelerate energy efficiency improvement as a fundamental contribution to economic and social development.

While significant efforts to achieve increased energy efficiency are essential for both industrialized countries and the developing world (including the formerly planned economies of Eastern Europe and the former Soviet Union), I concentrate my attention on the developing world in this paper. The focus on the developing world and formerly planned economies is appropriate, as it is in these regions that the greatest challenge will occur. Furthermore, without first-rate technical assistance on energy efficiency, the developing world will find it difficult indeed to meet the energy and environmental challenges of the future.

2 Patterns of Energy Consumption in Developing Countries

(a) Growth in energy demand

On a per capita basis, developing countries in 1991 consumed an average of 20 million British thermal units (Btu) in commercial fuels and an additional 8 million Btu in biofuels, compared with about 150 million Btu in Western Europe and more than 300 million Btu in the United States. Total primary commercial energy consumption in 1991 was 83 quads in the developing world and 64 quads in the former USSR and Eastern Europe, compared with 77 quads in the US and 88 quads in the rest of the OECD countries.[1] Among the developing nations, China accounted for 33% of the total, Asia (excluding China) 26%, Latin America 22% and Africa 11%.

Although total energy consumption in the developing world is less than half that of the developed world, and per capita consumption less than 15%, the rate of growth in energy consumption has been far higher in the developing countries for the last two decades (Figure 1). As a result, the share of the developing countries in world commercial energy consumption grew from 14% in 1973 to 27% in 1991 (Figure 2). Energy demand growth in the developing countries averaged about 5% per year during this period, compared with 0.6% per year in the United States and 1% per year for other OECD countries. Reasons for the rapid increase in primary energy consumption in the developing economies include:

- faster economic growth than in the OECD countries, particularly in the 1970s;
- migration from rural areas, where energy needs are met primarily with biomass, to urban areas where commercial fuels predominate;
- penetration of energy-intensive technologies (eg increasing use of fertilizers, personal vehicles and electric appliances);

- limited capability and resources to improve energy efficiency; and
- expansion of energy intensive industries.

(b) Energy consumption and GDP

Overall, commercial energy consumption[2] in the developing world grew around 20% faster than gross domestic product (GDP) between 1973 and 1987. The relationship between energy consumption and GDP has varied between regions and over time (Figure 3). In Asia (excluding China), energy consumption has grown slightly faster than GDP, especially since 1984. In China, which has experienced rapid GDP growth from the mid-1970s through the late-1980s, increases in energy consumption slowed markedly after 1979, mostly due to improvements in the industrial sector brought about by policies to encourage energy efficiency and major programs to allocate capital to energy efficiency investments (Levine and Liu, 1990). In Latin America, energy consumption has grown somewhat faster than GDP, especially during the recession of the early 1980s. In Africa, energy consumption has risen much faster than GDP, which has increased only slightly since 1980. Under-utilization of industrial capacity may have lowered efficiency in Latin America, while growth in the use of commercial fuels for residential purposes, such as cooking, has contributed to the observed trend in Africa.

(c) Sectoral energy consumption

For the developing countries as a group, industry accounted for 50% of final energy consumption in 1986, transportation for 22%, and buildings and agriculture for 29% (Table 1). The share of industry ranged from a high of 59% in China to a low of 30% in the Middle East. Transportation is nearly as important as industry in Latin America. Its relative importance is low only in China, where private transportation is much less used than elsewhere. Energy consumption in buildings and agriculture ranged in share from 36% in China, which has much more need for space heating, to 18% in the rest of Asia.

While industry is the major energy consuming sector in the developing countries, energy use in buildings and agriculture has grown faster (5.8% per year in 1973-1986) than industrial energy use (4.9%). Consumption in transportation averaged growth of 4.1% per year. Energy demand growth in buildings and agriculture was fastest in China and Africa, while it was about the same as growth in industry in Latin America and Asia. Transport was the fastest growing sector only in the Middle East. Growth in buildings energy use has been primarily due to increase in appliance ownership, switching from biomass to oil-based fuels for cooking, and construction of modern commercial buildings.

(d) Inefficiency of energy use

Overall, developing countries require substantially greater quantities of energy resources to deliver one unit of useful energy than do industrialized countries. One analyst estimates that the energy intensity (measured as commercial and non-commercial energy per unit GDP) of developing and Eastern European countries is approximately 60% greater than that of industrialized countries (Goldenberg, 1990).

It is, however, dangerous to infer very much about energy efficiency in developing countries from a comparison of their energy intensity with that of developed countries. Because exchange rates for developing country currencies are often misleading indications of the value of the currency—and other means of valuing local currencies are also uncertain—energy use per unit GDP is an especially uncertain indicator. Also, the structure of economies varies, both from one country to another and between developing and developed countries. Thus, even if energy/GDP ratios were accurate and meaningful, they would not necessarily be a good measure of energy efficiency for developing countries.

Nonetheless, it is certain that energy is used very inefficiently in developing countries (Levine, et al., 1990). The following examples for electricity production and use make this clear:

- The efficiency of generating electricity is often low in developing countries. Typical existing baseload power plants in OECD nations have heat rates (Btus of fuel per kilowatt hour) ranging from 9,000 to 11,000. New plants in developing countries are also often in this range. However, as power plants in developing countries age, deterioration in performance is accelerated by a lack of proper maintenance and spare parts, inadequately trained personnel and the use of low quality fuel. The result is that typical existing power plants often operate at 13,000 Btu per kilowatt hour, thereby increasing fuel requirements by 18 to 44%.
- Transmission and distribution (T & D) losses are very high in developing countries. A survey of 100 developing countries by the World Bank estimated average T & D losses at 17% (Escay, 1990). In OECD countries, typical T & D losses are in the range of 6 to 8%. Even correcting losses in developing countries for theft and inadequate billing procedures, it is likely that T & D losses in developing countries are at least 50% greater than those in OECD countries.
- The efficiency of use of electricity is often low. Typically, industrial processes—the largest users of electricity in most developing countries—are in the range of 30% less efficient than their counterparts in industrialized countries (Levine, et al., 1990; Gamba, et al., 1986; Goldemberg, et al., 1988). Studies of industrial energy and electricity savings opportunities in such countries as Thailand (Tectakeaw, P., 1988; US Agency for International Development, 1988), Indonesia (Chatab, I.N., 1988) and Egypt (Gelil, I.A., 1988) indicate reductions in process energy use of 15 to 30% with paybacks of one to three years. Comparable gains can be achieved in efficiency of electrical appliances. For example, a recent study in Indonesia suggests that new appliances could have 20 to 30% higher efficiencies than current levels with short paybacks, with much greater efficiency gains possible over time (Schipper, L. and Meyers, S., 1990).

These losses in the electrical system are compounded. If electricity generation and use are each 30% less efficient in developing than in developed countries, and transmission losses (not counting theft) are 16% instead of 8%, then twice as many energy resources are required to provide end-use service in a developing as in a developed country. Even though the price of electricity to a consumer is highly subsidized in most developing countries, the high economic cost of this power must be borne by the country. It is in many ways a great tragedy that the nations that can least afford to waste economic resources are often unable to avoid such losses, because of inadequate technology, infrastructures, human resources and essential facilities. Energy—because it is so essential to industrial and economic development, because it demands so much in the way of capital resources—is a prime example of this situation in developing countries.

(e) Electricity

Electricity use merits special attention because growth rates and attendant capital requirements have been especially dramatic in the developing world. Despite rapid growth in electricity supply, developing countries still use only 500 kWh of electricity per capita per year, compared with 5,000 kWh for Western Europe and 10,000 kWh for the United States. Power shortages present problems in many developing countries. Growth in electricity consumption in the developing countries has been higher (7.8% per year between 1973-1987) than the increase in overall final energy consumption, which averaged 5.1%. Increases in electricity consumption have been well above the economic growth rate. As incomes rise, the demand for amenities associated with electricity services (eg comfort, convenience and increased labor productivity) grows even faster.

3 Requirements for and Impacts of Energy Efficiency Improvements

(a) Requirements

Energy efficiency programs, policies and investments can have a major impact on the growth of energy demand; air pollution emissions felt at the local level (eg SO_x, NO_x, and particulates), regional level (eg acid rain), and globally (CO_2); and economic development in developing countries.

Figure 4 shows several recent scenarios of energy demand in developing countries through 2025. The Inter-governmental Panel on Climate Change (IPCC) scenario shows primary commercial energy demand in developing countries increasing by a factor of almost four over the 40 year period 1985 to 2025. The World Energy Conference moderate scenario (labeled WEC-M in Figure 4) and the Global Energy Efficiency Initiative (GEEI)-reference case both show energy use in developing countries increasing about threefold over 40 years.[3]

The scenarios labeled WEC-L and GEEI-initiative both represent cases in which substantial gains in energy efficiency take place over the next decades. Both of these cases show energy demand growth reduced by about 40% from the reference cases. This is accomplished through efficiency gains alone in the GEEI-initiative scenario, which makes explicit many of the requirements associated with this magnitude of reduction of energy growth. The saving, amounting to some 60 quads per year (75% of total US energy consumption today or about $400 billion per year at current energy prices), requires:

- Decisions by leaders of developing countries to place very high priority on energy efficiency. This means that energy efficiency must be equal to or greater in importance than increased energy supply; it also means that energy efficiency must be elevated on the economic and social development agenda.
- Similar decisions by leaders of international lending and economic assistance organizations to elevate the importance of energy efficiency. This will require changes in the practices of some existing international organizations as they relate to energy issues (Philipps, 1990).
- Decisions by the OECD nations to support large-scale assistance and lending programs for energy efficiency in developing countries.

It is possible that the negotiations to reduce growth in greenhouse gas emissions may provide a vehicle to initiate such large-scale assistance and loan programs. If so, it is important to have an idea of how much money will be required.

The GEEI scenarios show an average reduction in growth of about 1.5 quads per year between the reference and lower cases. This incremental saving begins at less than one quad per year and grows to more than two added quads of savings each year by 2025. Assuming a three year payback on energy efficiency investments[4] and a cost of energy of about $6 billion per quad, average investment of about $30 billion per year in new energy efficiency projects is needed. Less than $20 billion of investment funds would be needed in the early years, growing to more than $40 billion per year in the second decade of the next century. While these numbers are very rough approximations—being particularly sensitive to the ability of numerous institutions in the developing and developed world to execute efficiency projects on a very broad scale—they provide a rough measure of the magnitude of the capital requirements for energy efficiency in the developing world. Investments of such magnitude, properly executed, could lead to half of new energy demand being met by energy efficiency and half by new supply over the next four decades.

Of the $30 billion per year investment required, some fraction would come from:

- local sources in developing countries (public and private);
- the private sector of industrialized countries; and
- existing international lending agencies.

It is at this time not clear how much of this share each of these sources is able or likely to bear.

In addition to the loan requirements, assistance funds are needed so that the investment in energy efficiency is productively employed. How much is needed to make such an energy efficiency strategy possible? I do not know the answer.

However, two different approaches to thinking about the problem yield somewhat comparable results. The first approach is by analogy with the costs of spurring energy efficiency in advanced industrial countries, such as the United States, through utility demand-side programs, federal and state information programs (eg appliance labeling and state energy office activities, home weatherization funds and federal energy conservation R & D). The second approach is by building up and costing a set of activities—such as large-scale training and institution building—needed to support developing countries to create markets for energy efficiency. My best guesses, following both of these approaches, is that in the order of $1 to $2 billion per year of energy efficiency assistance is required for developing countries to create and administer energy efficiency policies and programs of the magnitude envisioned in this paper. This is on the order of 5% of the total required investments. It is important to note that these estimates, as well as those for capital requirements, are for developing countries alone. They do not include Eastern Europe or the former Soviet Union, which would require a separate analysis.

(b) Impacts

A comparison of the two cases (for either the GEEI or WEC scenario) shown in Figure 4 shows dramatic differences between them. The efficiency case reduces emissions of CO_2 from developing countries by about 25% (Levine, et al., 1990). Because these cases do not consider either fuel switching or increased growth of renewable energy systems, they do show continued growth in CO_2 emissions from energy production and use at about 2.5% per year. However, a long-term effort that reduced growth in coal use by substituting natural gas, renewable energy and/or nuclear power could reduce CO_2 emissions even further.

From the point of view of developing countries, the major benefit from such an efficiency program is the reduction in capital requirements. It is likely that average capital requirements for energy supply will be cut by about $50 to $60 billion—$25 to $30 billion per year in the short-term and $70 to $80 billion per year after 40 years. Thus, the net annual savings in investment, after the costs of efficiency investments (which I estimate to be about half the cost of new energy sources) are subtracted, is $20 to $30 billion per year. Making this magnitude of capital resources available for other development needs in the developing world would have a profound impact on the economic and social conditions of countries throughout the Third and Fourth Worlds. It is difficult to overstate the magnitude of impact that the freeing up of such funds could have. It is as if the entire resources of the World Bank were to more than double in the short-term, and then see accelerated growth over many decades.

From the point of view of the industrialized world, the increased efficiency in the developing world (combined with continuing gains in efficiency in the advanced economies as well) would yield very significant advantages: protection of the global environment, reduction of the stress on world oil resources, reductions in poverty in the Third World (and reductions in the national and regional conflicts that stem from such poverty) and improved world economic performance resulting from the saving of several tens of billions of dollars per year on energy expenditures without reducing energy services. While energy efficiency will not alone solve the problems of world poverty, global oil supply or the global environment, it can make a significant contribution to addressing these major issues. But it will not happen without the major commitments which I describe below.

4 Conclusions: An Agenda for Action

(a) The agenda for the industrialized countries

Global progress depends largely on actions by the higher income countries to set the pace (Helsinki Symposium, 1990). These actions include making tough policy decisions in order to realize potentials for efficiency improvement. Where global issues such as climate change are concerned, relatively small-percentage improvements in energy efficiencies in affluent countries can make more difference than relatively large-percentage improvements in poor countries. Moreover, the industrialized countries have more resources to invest in innovation and risk-taking.

In a similar vein, industrialized countries must provide substantial collaboration in a spirit of partnership. In many cases, substantial progress with efficiency improvement in lower income countries will depend on access to resources that in turn depend largely on the policies and actions of higher income countries. The main opportunities for initiatives on the part of industrialized countries are related to: training, to upgrade technical and managerial capabilities in developing countries; access to capital, to make financing available for cost-effective, energy efficiency improving investments through a combination of public and private sector sources; and support for institutional development, both nationally and internationally, related to a commitment to energy efficiency improvement.

(b) The agenda for the developing countries

The starting point for governing agencies in developing countries is for national leaders and energy sector decision makers to give efficiency improvement careful attention as a development investment option and to pursue it. Moreover, developing countries must improve institutional performance by building a management structure that is goal-oriented, adaptable and resilient. Efficiency improvements require effective indigenous institutional change (eg Wilbanks, 1990), a strategy for gap-filling and strengthening is urgently needed.

Improving incentives for energy efficiency through policy reforms in such areas as energy pricing and private sector roles will require facing challenges for policy reform. Those political economies that need reform the most are often the most resistant to it, because the required degree of painful adjustment is greater (Katzman et al., 1990). Regardless, we know from the experience of many countries that incentives such as market signals are a key to efficiency improvement (eg Hirst and Greene et al., 1981). Combined with effective institutions, in fact, they can stimulate powerful changes through dispersed voluntary decision making (eg Warford, 1989).

Entering the 1990s, with every energy and economic forecast pervaded by uncertainties, the nearest thing to a certainty about these alternative paths is that, in nearly every country, at nearly every level of development, in nearly every context, energy efficiency improvement needs to be a central component of an energy strategy for sustained development. But realizing the potential of efficiency improvement will call for forceful action by North and South alike, working together; and the longer we wait to act, the more serious the adverse impacts on developing economies and the global environment will be.

(c) Specific recommendations and their rationale

- Industrialized nations need to make clear the importance they place on increased energy efficiency, both for themselves and for the world. As a critical part of this commitment,]he industrialized nations should greatly increase technical assistance for energy efficiency to developing countries.
- This assistance is essential to develop the human resources and institutions in developing countries to be able to transform their energy economies.
- In steady state (eg after five years), technical assistance for energy efficiency from North to South of $1 to $2 billion per year could support the anticipated growth of energy services in the developing world and formerly planned economies while cutting the growth in demand for energy by more than 50%.[5]
- The present total development assistance from North to South is about $50 billion per year. Considering the importance of the energy sector to developing nations,[6] it is easy to justify 2 to 4% of assistance to developing countries for energy efficiency and perhaps a comparable amount for new supply.
- Indeed, this use of technical assistance from North to South is not only among the most beneficial uses of such funds for developing countries, it is also highly beneficial to the world economy and the global environment.
- Unfortunately, the actual amount of technical assistance for energy efficiency to developing nations is very small, a small fraction of what is needed to transform the energy economies. Because of the small size of the existing programs in industrialized countries to provide technical assistance to developing nations, it may be necessary to create new institutions within these countries to achieve this new mission.
- Initially, most of the $1 to $2 billion should be spent on training programs, strengthening institutions for energy efficiency in developing nations and energy efficiency pilot programs. The pilot programs can range from essential data gathering and policy development to demonstration projects and centers for energy efficient technologies.

- The industrialized world also needs to lead by example, demonstrating its commitment by improving its own energy efficiencies, investing in energy efficiency R & D, and otherwise supporting the continued advance and implementation of energy efficiency in the advanced economies.

- Developing nations need to make a strong commitment to energy efficiency as a major energy policy objective. This will typically mean a commitment to obtain half of the energy services needed in the future through energy efficiency gains.
  - For such a commitment to be achieved, technical and economic assistance from the North is essential. This assistance is needed to provide the initial impetus for large-scale energy efficiency activities in developing countries: trained staff; strengthened institutions for carrying out energy efficiency programs; feasibility studies to justify large capital investments in energy efficiency; a private sector that can supply energy efficiency equipment (including joint ventures with industrialized countries); and the full set of capabilities that will attract and effectively employ the large amount of capital for energy efficiency, most of which needs to come from the private sector (private financial institutions for most of the capital and private engineering, manufacturing and construction firms for the utilization of the capital).
  - The commitment in the developing countries must be made at the highest levels of government and the private sector. The government must be prepared to promote energy efficiency in a variety of ways: create new or strengthen existing institutions; craft national and local policies and programs for energy efficiency and implement them; make high quality personnel available to staff energy efficiency programs; and use the financial resources of the nation to increase capital availability for energy efficiency. The private sector must be prepared to invest funds and take risks to be a full partner in the creation of an energy efficient economy among the growing number of developing countries who are involving their private sector in providing energy services. The most important need will be for the public sector in developing countries to create an environment—through efficiency standards, fiscal mechanisms, industrial process modernization, energy pricing reform and the like—so that energy efficiency will happen routinely in the market place.
  - The developing countries need to make certain that the technical and economic assistance programs will be used to create the long-term capability of the country to achieve energy efficiency without external assistance. As such, the most important investment of these funds in early years needs to be in providing education and training to the individuals who will later serve as the leaders of the energy efficiency programs.
  - There needs to be put into place in all countries (both North and South) a system to measure the performance of energy efficiency in the economy. (Such a system could be coordinated, and partially funded, by one of several international agencies.)

- International agencies involved either in energy and environmental affairs (eg the United Nations, the International Energy Agency) or in large-scale lending to developing nations (eg the World Bank and the regional banks) need to reset their priorities to support energy efficiency.
  - The international banks could play a crucial role in promoting energy efficiency, but a significant change in policy is required.
  - If the North commits to providing $1 to $2 billion per year of technical assistance for energy efficiency to the South, this should serve (among other purposes) to greatly encourage a change in policy of the banks as well as the creation of energy efficiency loan funds to work along with the bilateral (or multilateral) efficiency assistance.
  - A total of $20 to $40 billion per year of capital investment for energy efficiency in developing nations and formerly planned economies is easily justified. With the training of technical and managerial personnel and the strengthening of institutions for energy efficiency, it should be possible within several years to use such energy efficiency investment funds productively.
  - There is a need for an institution to carry out a major worldwide energy efficiency campaign in both developing and industrialized countries. Among the responsibilities of such an institution is the management and coordination of all major international programs in energy efficiency assistance and loans for developing nations. At present, there are many different international institutions that are playing some role relating to energy efficiency, but all have other objectives that are more important than energy efficiency to them.[7] Additionally, there will, under this proposal, become a critical need for coordination among the bilateral (or possibly multilateral) programs supporting energy efficiency.
  - An international organization whose sole mission involved promoting energy efficiency worldwide could over time evolve from a coordination role to one of overseeing the implementation of some programs. For example, such an organization could take training funds for energy efficiency from many different industrialized countries and make certain that an organized, non-duplicative training program was created. It could also organize extensive feasibility studies of energy efficiency projects, and then assist in raising capital to fund the most promising of the projects.
- There is a need to figure out how to overcome the barriers to a concerted international effort to enhance energy efficiency. If two or three leading industrialized nations (eg the United States, Japan and Germany) agreed to commit a total of $1 billion per year to this effort, and began the complex process of figuring out how the money would be spent, by and with whom, for what specific purposes, as well as a (likely interim) structure for coordination of the energy efficiency programs, it would be possible to create such a new, global initiative. In the absence of such specific actions by these countries, we will continue to talk about these great ideas but will likely see little action.

Note: This article is a revised version of an earlier paper entitled 'The Contribution of Energy Efficiency to Sustainable Development in Developing Countries.' The earlier paper was co-authored with Stephen Meyers and published in the Natural Resources Forum special issue on the occasion of the United Nations Conference on the Environment and Development, Volume 16, Number 1, February 1992.

Footnotes

1 The source of energy data in this section is the International Energy Agency, World Energy Statistics and Balances 1971-1987, Paris, 1989. Primary energy consumption includes all non-energy uses; final energy consumption does not include petroleum products used for non-energy purposes, but does include feedstocks for the petrochemical industry. Consumption does not include biomass, use of which is not well-documented. Within the developing world, China accounted for 36% of total consumption, Latin America for 24%, Asia (excluding China) for 20%, Africa for 12% and the Middle East for 9%. We have not included as developing countries a number of countries with relatively high per capita GNP that have often been classified as part of the developing world. These countries are Hong Kong, Singapore, Taiwan, South Korea, Israel, Kuwait, Oman, Qatar, Saudi Arabia and United Arab Emirates. OECD is the Organization for Economic Cooperation and Development and includes Canada, the United States, Western Europe, Japan, Australia and New Zealand.

2 Through the text we will use the expressions 'commercial energy consumption' and 'energy consumption' interchangeably. The expressions exclude traditional (biomass) fuels. If biomass is included, this will be noted explicitly.

3 The descriptions of the scenarios in Figure 4, including the assumptions underlying them, are found in World Energy Conference (1990), IPCC (1990) and Levine, et al., (1990). For purposes of comparison, we have extrapolated the results of the World Energy Conference Scenarios to 2025.

4 This may be a somewhat pessimistic assumption. Experience in the United States has shown a substantial potential for energy savings with paybacks of two years and less. However, considering the virtual non-existence of industries to support energy efficiency projects in developing countries and the costs of establishing such activities throughout an economy, we believe this is a reasonable assumption. Note that the assumption of a two year payback would reduce the capital requirements to achieve the same level of energy savings as shown in the scenario.

5 This $1 to $2 billion per year of technical assistance is in addition to—and essential for the success of—the annual $20 to $40 billion of capital investment for energy efficiency projects which needs to come from private and public sector sources.

6 The energy sector typically consumes 20 to 40% of total public investment in developing countries (ASAID, 1988).

7 At present, the most important players among the international institutions are the Global Environmental Facility (a joint World Bank/UNEP activity), the World Bank, the regional banks (eg the Asia Development Bank), the United Nations (especially the UN Environment Program (UNEP) and the UN Development Program (UNDP)) and (for data gathering) the International Energy Agency. In addition, the

organizations that are being or will be created or empowered to carry out the United Nations Conference on Environment and Development will play an important role in energy efficiency programs for developing countries. Any new organization with a specific mandate for energy efficiency will need to work with these existing entities. Ideally, an effort would be made to reduce rather than increase the number of players, so that coordination does not consume the effort.

References

Chatab, I.N., 1988. 'Current Energy Conservation Status in Indonesia', in *Energy Conservation* (Proceedings of the International Energy Conservation Symposium, October 1988, Islamabad, Pakistan).

Gamba, J.R. et al., 1986. 'Industrial Energy Rationalization in Developing Countries', World Bank, Johns Hopkins University Press.

Gelil, I.A., 1988. 'Egypt's National Energy Conservation Strategy' in *Energy Conservation* (Proceedings of the International Energy Conservation Symposium, October 1988, Islamabad, Pakistan).

Goldemberg, J., 1990. 'Policy Responses to Global Warming', in J. Leggett, ed., *Global Warming: The Greenpeace Report*, Oxford: Oxford University Press.

Goldemberg, J., T.B. Johansson, A.K.N. Reddy, and R.H. Williams, 1988. *Energy for a Sustainable World*, New Delhi: Wiley Eastern Ltd.

Helsinki Symposium, 1990. 'Energy and Electricity Supply and Demand: Implications for the Global Environment', (Issue paper, Senior Expert Symposium on Electricity and the Environment (Helsinki, May 1990), Vienna, October 1990).

Hirst and Greene et al., 1981. *Energy Use from 1973 to 1980: The Role of Improved Energy Efficiency*, Oak Ridge, Tennessee, Oak Ridge National Laboratory ORNL-CON-79.

Katzman, H., L. Hill, M. Levine, and T. Wilbanks, 1990. 'The Prospects for Energy Efficiency Improvements in Developing Countries', *The Energy Journal*, forthcoming.

Levine, M., A. Gadgil, S. Meyers, J. Sathaye, J. Stafurik, and T. Wilbanks, 1990. 'Assessment of a Global Energy Efficiency Initiative: Draft Report', November 1990.

Levine, M.D., and X. Liu, 1990. 'Energy Conservation Programs in the People's Republic of China', Lawrence Berkeley Laboratory, Berkeley, California.

Philips, M., 1990. 'Alternative Roles for the Energy Sector Management Assistance Program in End-Use Energy Efficiency', (draft), International Institute for Energy Conservation (IIEC), October 1990.

Schipper, L. and S. Meyers, 1990. 'Improving Appliance Efficiency in Indonesia'. *Energy Policy*, forthcoming.

Tectakeaw, P., 1988. 'Energy Conservation in Thailand' in *Energy Conservation*, Proceedings of the International Energy Conservation Symposium, October 1988, Islamabad, Pakistan.

US Agency for International Development, 1988. *Power Shortages in Developing Countries: Magnitude, Impacts, Solutions, and the Role of the Private Sector*, Washington, DC, March 1988.

US Agency for International Development, 1988. 'The National Energy Savings Goal for Industry, 1986-1991', prepared for the Government of Thailand by RCG/Haggler, Bailly, Inc., March 1988.

Warford, J.J., 1989. 'Environmental Management and Economic Policy in Developing Countries', in G. Schramm and J. Warford, eds., *Environmental Management and Economic Development*, Johns Hopkins University Press, pp. 7–22.

Wilbanks, T., 1988. 'Impacts of Energy Development and Use, 1888-2088', in *Earth '88*, (Proceedings of the National Geographic Society Centennial Symposium, Washington, DC, pp. 94-144).

Wilbanks, T., 1990. 'Institutional Issues in Energy R & D Strategies for Developing Countries', prepared for Workshop on Energy Research and Development for Developing Countries, Oak Ridge, TN, November 1990.

Figure 1
Average growth rates, primary energy consumption

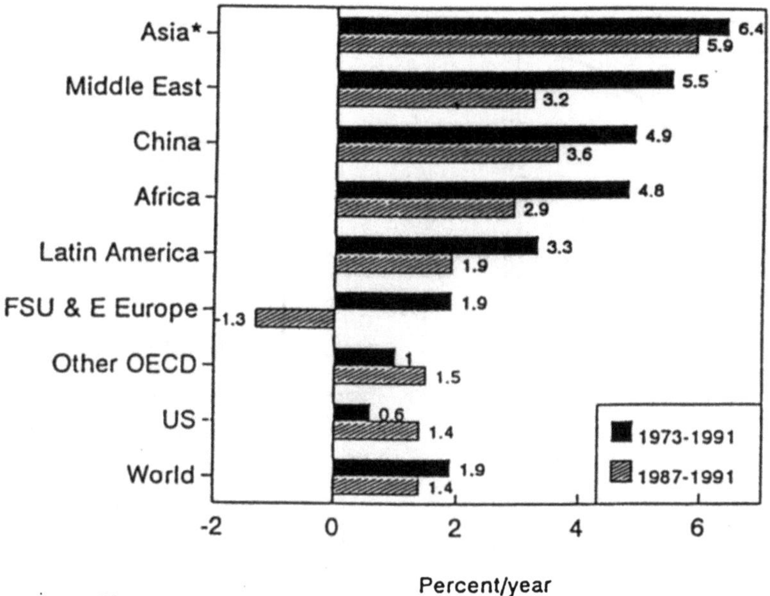

Source: BP, 1992. * Excluding China & Japan.

Figure 2
World primary energy consumption, 1973-91

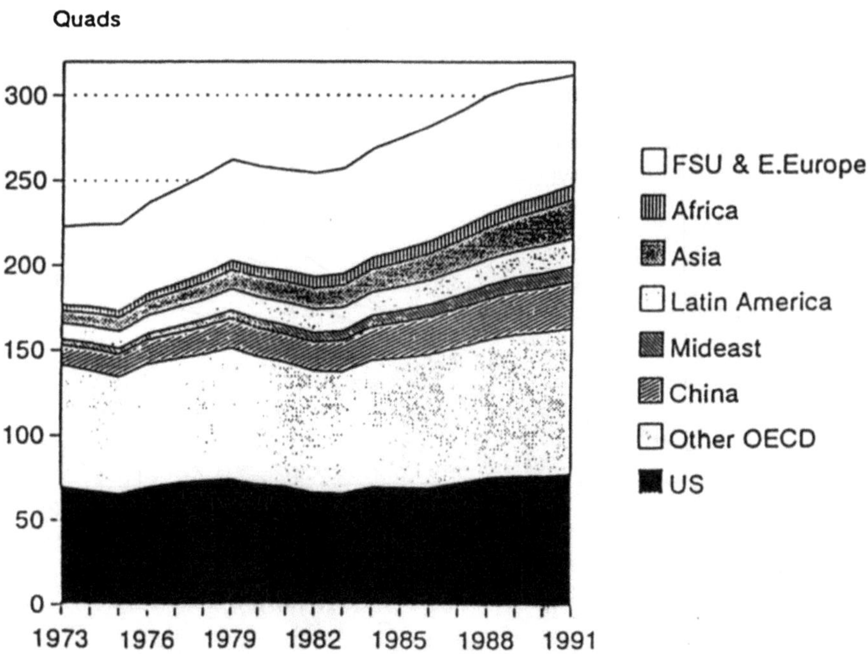

Figure 3
Energy and GDP indices

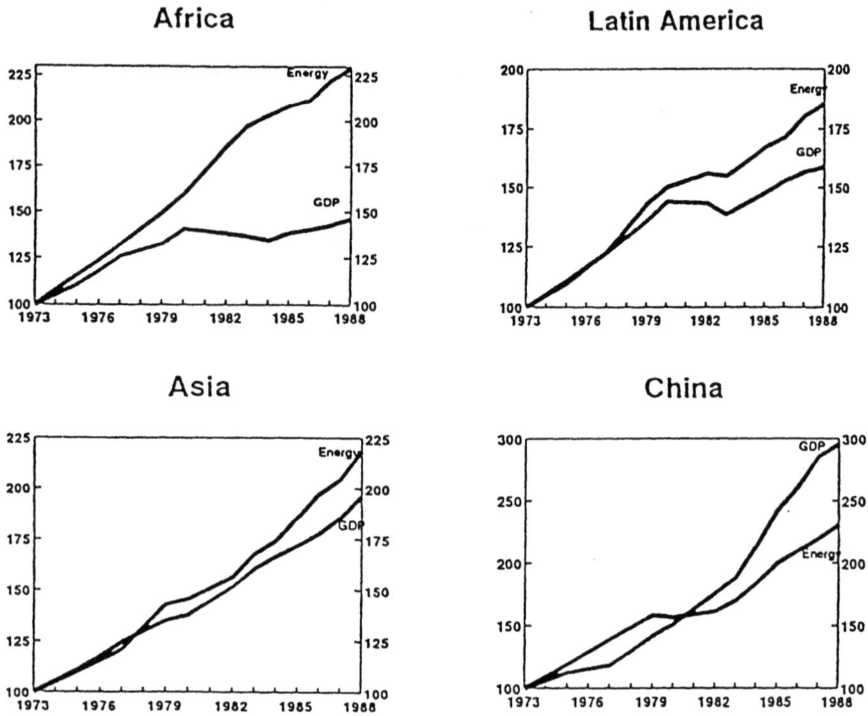

Figure 4
Energy scenarios for developing countries, 1985-2025

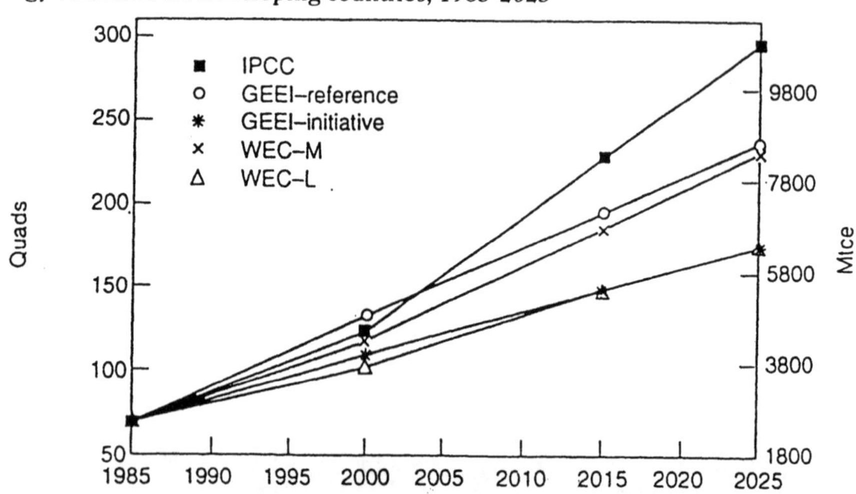

WEC 2020 projections are extrapolated to 2025. All WEC projections are adjusted

Table 1. Sectoral Shares in Final Energy Consumption in 1986 (%)

Region	Industry	Transportation	Building & Agriculture
Latin America	39	38	23
Asia	55	27	18
China	59	6	36
Africa	45	32	23
Middle East	30	34	36
Total Developing Countries	50	22	29

II.3 Technology for Large Scale Utilization of Photovoltaic System

K. Yamada

Global Environment Laboratory, Faculty of Engineering, University of Tokyo, Japan

Abstract

In order to realize sustainable development which is in harmony with the global environment, it will be necessary to change the present world energy system.

The total global insolation energy available is 10,000 times the present world energy consumption. The amount of solar energy insolated on the earth is sufficient to meet the world's energy demands, but its density is low.

This paper considers various forms of renewable energy, such as photovoltaic (PV), solar, thermal, wind, wave and biomass energies, presently under development and discusses the possibilities of applying some of them.

PV Energy Systems

One very important factor in determining the energy pay back time (EPT) and the economics of PV energy systems is the energy conversion efficiency of solar cells. The conversion efficiency of solar cells has been steadily increasing, and is expected to improve still further (see Figure 1). Both amorphous silicon (a-Si) and polycrystalline silicon (poly-Si) are possibilities for future large-scale application. The costs of large-scale PV energy systems should be low, EPTs brief, and life expectancy long.

The EPTs and costs for a-Si and poly-Si have been evaluated. The evaluated scales of annual PV cell production are 0.01, 1 and 100 gigawatts per year. The process flow diagrams of a-Si and poly-Si cells are shown in Figures 2 and 3. Table 1 shows the assumptions for the evaluation. It is assumed that the PV systems are installed in Japan and that they can be connected directly to power grids, without using battery systems. The results of the calculations are shown in Table 2.

In the assessment of the annual production scale of 0.01 gigawatts (Case 1), the current most advanced technologies are assumed. The first target of the EPT seems to be 3 years, and the cost 150 Yen per week. EPTs and costs decrease considerably with increases in production. The EPTs of the systems using poly-Si cells are 5.76, 3.59 and 3.26 years for Cases 1, 2 and 3 respectively. The EPTs of the systems using a-Si cells are 6.35, 3.59 and 3.00 years. The EPT valuers for both the poly-Si and a-Si systems almost reach the target after Case 2 in a relatively short space of time. The costs of the systems using poly-Si cells are 547, 314 and 233 Yen per week for Cases 1, 2 and 3, while the costs for systems using a-Si cells are 790, 288 and 210 Yen per week respectively.

After Case 2, a system using a-Si cells becomes more economical than those using poly-Si cells. However, in Case 3, the cost of the system using a-Si is higher than the first target.

In order to reduce the electricity cost by PV systems, two cases were investigated. The first involved the roof-top type (or roof-type) systems in Tokyo which use less supporting material than Cases 1 to 3. The second involved the PV systems installed in the Gibson Desert in Australia where the solar energy is 1.8 times higher than it is in Tokyo. The results are shown in Tables 4 and 5.

The costs of the roof-type systems at the cell production scale of 1 gigawatt per year can be reduced from 288 to 173 Yen per week by applying a simple supporter rack and high inverter conversion efficiency. The EPT systems can also be reduced from 3.54 to 0.64 per year.

When the total annual expenditure is assumed to be 15% of the investment cost, the electricity cost is calculated to be 26 Yen per kilowatt hour. This electricity cost is approximately the same as the commercial price in Tokyo. Roof-type PV energy systems can supply about 8% of Tokyo's total demand.

On the other hand, because of the high cost of batteries, the cost of generating electricity with highly incident solar energy in the Gibson Desert is a high 511 Yen per week. Without batteries the cost decreases to 102 Yen per week (20 Yen per kWh)—lower than the first target.

These results suggest that:

(1) Roof-type PV energy systems using a-Si cells can be economically applied on a large scale, preferably in high insolation areas.
(2) Because of the high cost of batteries, centralized PV energy systems generate very costly electricity. For the large-scale application of the system, it will be necessary either to reduce the battery cost considerably or to develop an energy application system which does not require any battery back-up.
(3) In the near future, PV energy systems will be very useful for the private sector in many countries, including developing countries. It may take a long time, however, to realize centralized PV energy systems.

Methane Synthesis by the Co-Processing of Biomass and Fossil Fuel

Atmospheric CO_2 can be fixed as ligno-cellulose by photosynthesis as shown by Example 1:

(1) $CO_2 + 0.72H_2O$ ==> CH1.44 O0.66 + $1.03O_2$

Biomass is not stable over a long period of time. In order to prevent the biomass carbon from re-entering the atmosphere as CO_2 from biomass decay, the biomass should be processed.

Steinberg[1] proposed the Hydrocarb process which produces methanol and carbon as shown by Examples 2 to 4. The co-processing of biomass is illustrated with coal (Example 2), oil (Example 3), and natural gas (Example 4).

(2) CH1.44 O0.66 + 3CH0.8 O0.1 ==> 3.04C + $0.69CH_3OH$

(3) CH1.44 O0.66 + 0.66CH1.7 ==> C + $0.66CH_3OH$

(4) CH1.44 O0.66 + 0.3CH4 ==> 0.4C + $0.66CH_3OH$

In each co-processing system, carbon (which can be either stored or used as a fuel) and methanol as a liquid fuel are produced. If the carbon is sequestered in the earth and methanol is utilized as a fuel, the net quantity difference between the biomass carbon used as a raw material and the resultant methanol is the total CO_2 removed from the atmosphere. Biomass is produced by the photosynthetic fixation of CO_2, which is an energy free process. The efficiency of energy utilization (EEU) can be calculated by Example 5.

$$(5)\quad \text{EEU} = \frac{\text{Heat of combustion of methanol}}{\text{Heat of combustion of fossil fuel}}$$

The calculated values of EEU and the amount of CO_2 removed are shown in Table 6.

The Hydrocarb process may be economically feasible under specific circumstances. However, technological difficulties exist in the heat recovery step at high temperatures, and the continuous operation under high pressure and dry conditions in the hydrogenation process. In the Hydrocarb process, the hydrogenation reaction of biomass by dry H2 can proceed only under severe conditions such as 1100°C, 50 atm. However, if a water/gas reaction is applied for biomass, the hydrogenation occurs at 1000°C under atmospheric pressure of H2, CO, CH4 and CO_2 gases, and carbon can be produced.

We named the process the biomass-carbon process. It is shown in Figure 3 at the annual production of 1 Mtoe for methanol. By this process, CO_2 zero emission methanol can be obtained when the carbon mole number of raw material biomass and that of product methanol are the same. The combustion heat of methanol produced is 1.79 times higher than that of natural gas. The excess heat of 79% comes from the biomass and corresponds with 32% of its combustion heat. The remaining 68% heat content of the biomass is used for heating in the biomass-carbon process. Because of the mild conditions, this process is much more easily industrialized than the Hydrocarb process.

An economic evaluation was done for the biomass-carbon process which produces 1 Mtoe per year of methanol. The raw material and product costs used are the same as reported elsewhere.[2] The investment cost for the main process is calculated to be 100 billion Yen, with another 100 billion Yen for utilities and infrastructure. The annual expenditure for labour, maintenance and tax is assumed to be 40 billion Yen (20% of investment cost). The total balance yields an annual deficit of 31.9 billion Yen. This deficit corresponds with 40500 Yen/t-C which is comparable with the values 26900 Yen ~ 39500/t-C for the separation and sequestration of CO_2 into the deep ocean[3] and the value of 34000 Yen/t-C by the charcoal method in Brazil.[4] These results indicate that the co-processing method involving CO_2 can produce clean fuel, and that developing countries seem suitable for the plant site from the viewpoint of raw material supply.

Conclusion

PV energy systems and the co-processing method of biomass and fossil fuels were evaluated in order to investigate the possibility of renewable energy technologies.

It will be possible to use roof-type PV systems in the near future. There are difficulties in large-scale application of PV systems with battery back-up. The co-processing method is an interesting possibility for future application.

Acknowledgments

The author gratefully acknowledges the contribution of the members of the CO_2 Group of the Japanese Chemical Engineering Society who conducted the evaluation work on PV energy systems.

Footnotes

1 Steinberg, M, 1991. 'Biomass and Hydrocarb Technology For Removal of Atmospheric CO_2.' BNL 4410R, February 1991.

2 Yamada, K., et al, 1993. Methanol and Carbon Production by the Coprocessing of Biomass and Fossil Fuel, proceedings 9th Energy Conference, Tokyo, January 1993.

3 Ishitani, K., et al, 1991. Energy Shigen Gakkai-Shi 12 (3), p. 66.

4 Yamada, K., et al, 1993 ibid

Figure 1
Change of conversion efficiency in solar cell

③ PV energy systems are very useful for private sectors of many countries including developing countries in the near future.
It may take a long time to realize centralized large PV energy systems.

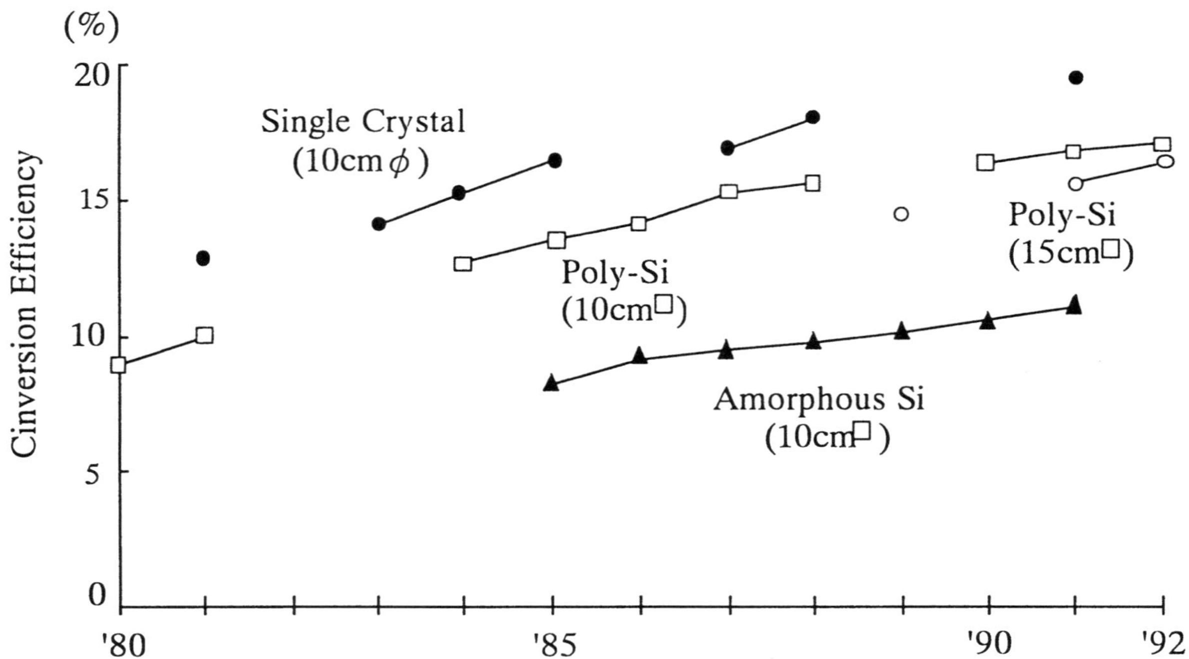

Figure 2
Process flow of the poly-Si solar cell production

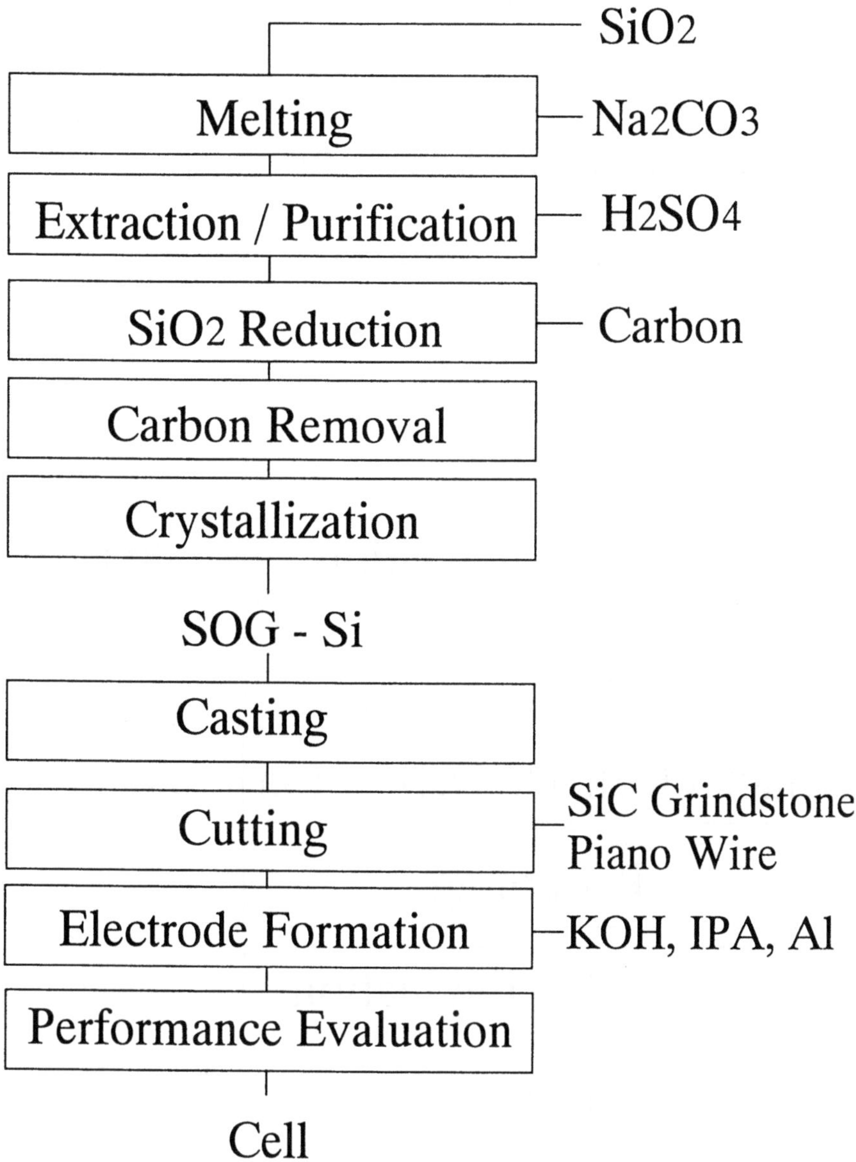

Figure 3
Process flow of the amorphous-Si solar cell production

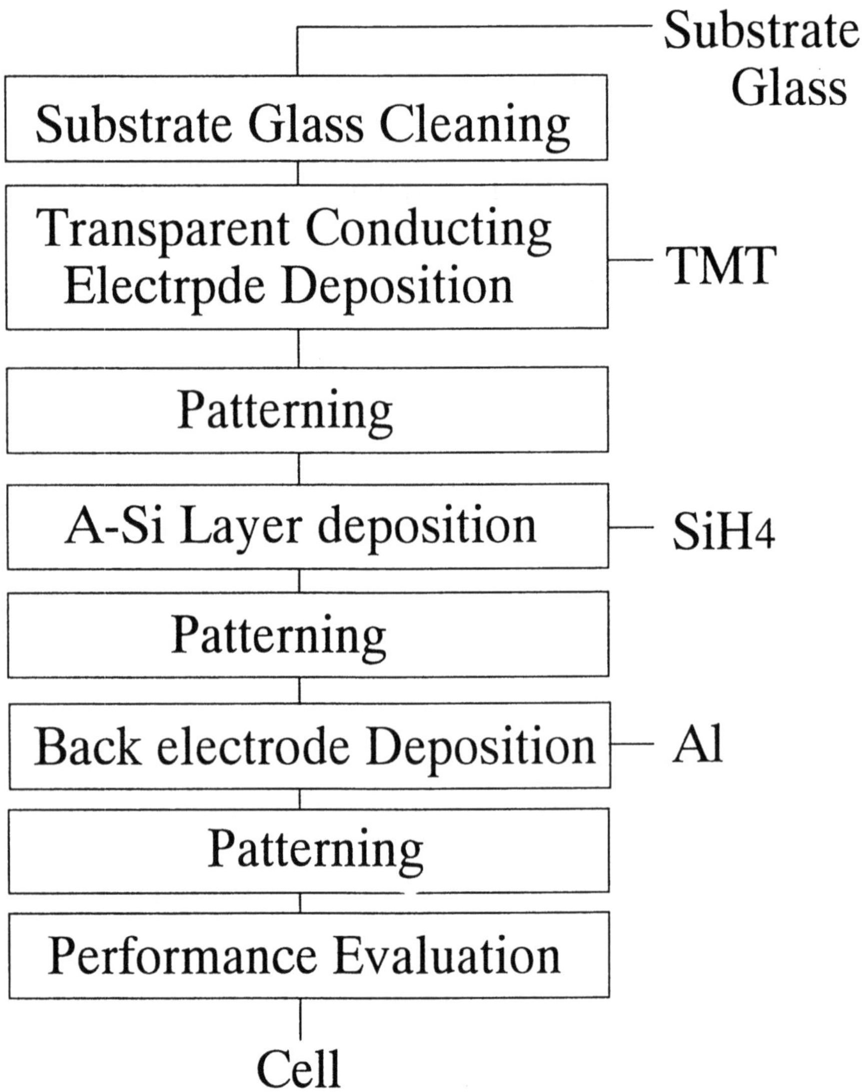

Figure 4
Methanol and carbon production from wood and natural gas (biocarbon—WG process)

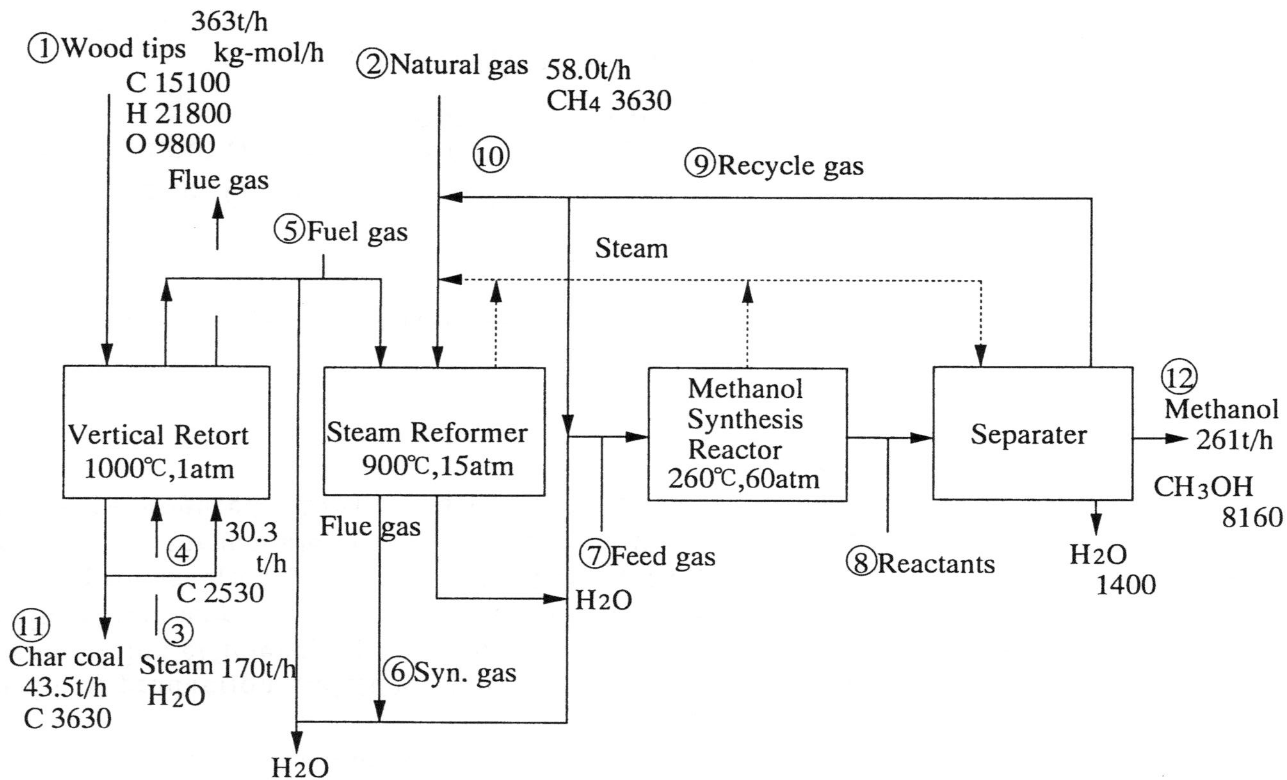

Table 1
Assumptions and calculated values for the PV systems

	Poly-Si			Amorphous-Si		
Annual cell production scale[GW/y]	0.01	1.0	100	0.01	1.0	100
Cell conversion efficiency[-]	0.15	0.17	0.20	0.08	0.13	0.16
Cell dimension[mxm]	0.1x0.1	0.15x0.15	0.2x0.2	0.3x0.4	0.4x1.2	0.4x1.2
Cell area produced[m2/y]	$6.94x10^4$	$6.13x10^6$	$5.21x10^8$	$1.25x10^5$	$7.69x10^6$	$6.25x10^8$
Package factor for module[-]	0.82	0.86	0.88	0.89	0.92	0.92
Module Area produced[m2/y]	$8.47x10^4$	$7.12x10^6$	$5.92x10^8$	$1.40x10^5$	$8.36x10^6$	$6.79x10^8$

Table 2
EPTS and cost of PV energy systems using poly-Si cells

Case	1		2		3	
Production seale (GW)	0.01		1.0		100	
Electricity output (kWh/y)	$8.64x10^6$		$8.64x10^8$		$8.64x10^{10}$	
	EPT (y)	cost (¥/W)	EPT (y)	cost (¥/W)	EPT (y)	cost (¥/W)
Module production	2.83	344	1.29	158	1.01	109
BOS production	2.93	203	2.21	156	2.25	124
Total	5.76	547	3.50	314	3.26	233

[BOS (Balance of system); supporting rack, inverter etc]

Table 3
EPTs and costs of PV energy systems using a-Si cells

Case	1		2		3	
Production seale (GW)	0.01		1.0		100	
Electricity output (kWh/y)	$8.64x10^6$		$8.64x10^8$		$8.64x10^{10}$	
	EPT (y)	cost (¥/W)	EPT (y)	cost (¥/W)	EPT (y)	cost (¥/W)
Module production	1.56	476	0.60	107	0.43	69
BOS production	4.79	314	2.99	181	2.57	141
Total	6.35	790	3.59	288	3.00	210

Table 4
EPTs and costs of PV energy systems using a-Si cells applied for detached house roofs in Tokyo

Module production scale (GW/y)	1	100
Roof area (m^2)	$42.8x10^6$	$42.8x10^6$
Electricity output		
(GW)	5.12	6.30
(GWh/y)	5130	6310
EPT(y)		
Module	0.50	0.37
BOS	0.14	0.12
Total	0.64	0.59
Cost (¥/W)		
Module	107	69
BOS	66	53
Total	173	122
Cost (¥/kWh) at indirect ratio of 15%		
Module	16	10
BOS	10	8
Total	26	18

Table 5
EPTs and costs of energy systems using poly-Si cells applied for power stations in Australia

Module production seale (GW)	1	100
Plant area (km^2)	9.8	860
Electricity output		
(GW)	1.4	145
(GWh/y)	$1.15x10^3$	$1.10x10^5$
EPT (y)		
Module, rack, inverter	1.73	1.30
Battery	0.64	0.71
Others	0.12	0.08
Total	2.49	2.09
Cost (¥/W)		
Module, rack, inverter	119	91
Battery	669	409
Others	4	11
Total	792	511
Cost (¥/kWh) at indirect ratio of 15%		
Module, rack, inverter	22	18
Battery	122	81
Others	1	2
Total	145	101

Table 6
Efficiency of energy utilization (EEU) and CO_2 removed

Case	EEU (%)	Net CO_2 removed from the atmosphere (g-C/Mcal-methanol)
Biomass-Coal	42	3.3
Biomass-Oil	110	40
Biomass-Natural gas	176	40

The Hydrocarb process may be economically feasible under specific circumstances. However, technological difficulties exist in the heat recovery

II.4 Possibility and Evaluation of CO_2 Disposal in the Deep Ocean

H. Ishitani and R. Matuhashi

Abstract

Among measures for the reduction of CO_2 emission from fossil fuel use, the deep ocean sequestering of CO_2 is now being considered as one of the alternatives which could be technically feasible within a few decades. There are still some technical uncertainties, some unknown ocean environmental factors and a question concerning the time necessary to isolate waste CO_2 from the atmosphere. But sequestering is regarded as one of the best and most realistic measures against global climate change because of the deep ocean's capacity to absorb CO_2.

There are a number of possibilities concerning disposal, recovery and separation systems for flue gases, and concerning final dumping systems. This paper investigates these possibilities, makes rough estimates of the cost of such systems, and compares the results with a Netherlands study.

This presentation discusses briefly basic concepts of the disposal system, estimates of costs, especially for energy, for the various systems, technical availability, and some other issues such as an evaluation of environmental effects on the ocean bed.

1 Objective

Possible countermeasures to global warming problems from CO_2 emission:

(1) Conservation of energy—from increase in efficiency; temperature rise in power plant; suppression of unnecessary energy use, etc.

(2) Switching from fossil fuel to non-carbonated energy resources, such as solar energy, or to other alternative fuels with less carbon.

(3) Recovery and isolation of CO_2 from flue gas emanating from fossil fuel burning.

(4) Fixation of CO_2 from the atmosphere (afforestation, etc).

None of these measures is capable of solving the problem alone. Therefore, it is necessary to try to utilize all available means simultaneously and, at the same time, to keep studying any technologies which might be able to contribute towards a solution. Until now, the technologies in category (3) above have had a poor technological basis. What is needed is a feasibility study of the technical possibilities.

2 The Ocean Sequestering of CO_2—Background and Present Situation

The sequestering of CO_2 into the ocean was originally proposed by Dr M Steinberg of Brookheaven National Laboratory. From a technical point of view, it seems to be the most practical and effective short-term solution. Most electricity companies are exploring this technique in relation to heavy industry companies, with research ranging from basic research on CO_2 behaviour under high pressure to field experiments on pilot plants for CO_2 recovery. Many research groups in AIST have also been working on basic studies of CO_2 characteristics and of recovery and disposal plants.

CRIEPI began with some enthusiasm a basic study of CO_2 characteristics, and conducted some offshore experiments on the deep sea disposal of dry ice. It is intended to hold a symposium 'The interaction of the ocean and CO_2' to discuss the biological and physical effects of CO_2 in the deep ocean. The Mitsubishi Heavy Industry Company, which first proposed such disposal systems, will next week present a paper on the technical feasibility of a system which disposes of CO_2 into the deep sea, to the Japan Naval Arch. Conference. Two years ago, as a first step, we began an attempt to evaluate from a technical point of view the feasibility of CO_2 removal and disposal, although we were aware of the fact that there were still many aspects of CO_2 behaviour which were unknown.

3 The Basic Concept

3.1 The overall system and its efficiency

A CO_2 disposal system consists of two categories—a CO_2 removal and recovery system, and a CO_2 disposal system.

3.2 Disposal system components

a) Recovery and solidification or liquefaction.
b) Collection to the land basement; storage on the seashore site.
c) Transportation to the injection site.
d) Disposal site equipment.
e) Pumping and injection.
f) CO_2 behaviour after injection.

There are three options:

- Pump down from the seashore base via a seabed pipeline (Steinberg).
- Pump down from an offshore station, to which collected CO_2 is transported by shuttle tankers for liquefied or solidified CO_2 (Mitsubishi).
- Drop solidified CO_2 (dry ice) from a dump conveyor ship (Echigo).

The first two options were studied in detail and compared, taking into account the actual possible flow rate of CO_2 recovered from all fossil electricity plants in the Kanto area.

3.3 Basic technical issues for such systems

The conditions of CO_2 disposal to the ocean:

- The density of liquefied CO_2 or dissolved seawater with CO_2 should be larger than that of the surrounding water.
- To select the point where the time taken for the dumped CO_2 to return to the atmosphere should be 1,000 years.
- If CO_2 remains for a long time at some particular points in the ocean, to identify those places which have sufficient capacity and which will carry the least risk of damage to the surrounding biological and physical environment.
- An attempt should be made to minimize the total necessary consumption.

Therefore we should dump CO_2 to a depth of 3,000 meters with sufficient pressure to ensure that the density of the CO_2 or of the seawater dissolving the CO_2 should be larger than that of the surrounding seawater.

3.4 Comparison of the pipeline disposal system with the offshore station system

The main problems of the simpler pipeline disposal system are:

(1) Except for very specialized cases (such as Kanto 120km), the system needs a long, wide pipeline with only a slight inclination.

(2) There are technical difficulties in the construction and maintenance of a highly pressurized underwater pipeline with an open outlet to a depth of over 3,000 meters.

(3) There is a problem in the formation of clathrate, causing choking at the outlet.

(4) It is not easy to move or to control the injection point after construction. There are also difficulties in protecting a wide pipeline at the seashore from ships or from fisheries operations.

(5) The difficulties increase for injection points at still deeper levels, where the environmental advantages may be greater.

On the other hand, the argument in favour of this system is that it injects CO_2 automatically from an offshore storage base without any difficult operation or handling.

3.5 The basic concept component of an offshore disposal system

(1) A collection pipe from each power unit to the seashore base storage tank. Liquefaction may be necessary in order to decrease the capacity of the CO_2 pipe. The storage tank is necessary as a buffer between CO_2 generation and transportation by a CO_2 tanker.
(2) A shuttle CO_2 tanker to transport CO_2 between the seashore base and the offshore disposal station.
(3) An offshore disposal station. This involves the positioning of the station, a vertical injection pipe to the bottom of the ocean (sometimes 10,000 meter depth), and equipment for off-loading the CO_2 from the tanker.

4 A Case Study

Target area and plants

An estimation of the total amount of CO_2 generated in fossil fuel power plants in an electric power company in the Kanto area.

CO_2 emission by fossil fuel power plants in the total Kanto area:
80.3 CO_2-Mt (24.4 C-Mt) per year
2.54CO_2-t/s = 2.21m3(300atm) per sec
Compression energy to 70atm for 1 gigawatt ==>690MW:
2.3% in primary energy equivalent (estimated as 1.44 x theoretical value)
Necessary storage tank capacity:
450Kt of CO_2
Necessary tanker:
150Kt x 2 tankers with 70 atm liquefied CO_2
Stay 8 hours:
loading speed 2.8t per sec. Cruise 120km within 8 hours at 8 knots in normal operation.
Flow rate at injection points:
2.5CO_2-t/s for 1 meter diameter outlet, 3 meter per second flow rate.

5 Environmental Effect Around Injection Points

The most important remaining significant problem to be considered in the realization of such a system is that of environmental evaluation.

The physical and chemical characteristics of CO_2 at low temperature and high pressure are not clear, which is why there have been many recent attempts to clarify the issue.

For an evaluation of the effect of CO_2 disposal on the affected area, by a very simple macroscopic model using the information available, the following assumptions were made:

- A simple macroscopic diffusion equation using conventional known data was assumed.

- Clathrate formation was not considered.
- A constant and stable downward flow from the pipe end was assumed.
- No consideration was given to tidal flow at the sea floor.
- Calculation of PH change was based on major chemical equilibrium equation.

Using these very simplified assumptions, it is already clear that such a disposal will have an effect on some of the surrounding area. More detailed evaluation is needed, and more information required on the effects of PH change and of the CO_2 pool on sea floor creatures in the affected area.

6 Summary and Remaining Problems

Although there are still many unknown factors concerning CO_2 characteristics and the technical feasibility of each element of the disposal system, the feasibility of the individual elements and their technological problems are summarized in Table 3.

Further summary observations are as follows:

(1) Although it is necessary to improve the rate of recovery on the land, the technological feasibility is sufficiently high.

(2) The systems from liquefaction/compression to sea dumping are simple. As CO_2 is not essentially dangerous, technological issues are not significant, except for the problem of location and the necessity to develop the technology to lay pipes on the ocean floor at a depth of over 3,000 meters.

(3) Except for the problem of operation, the offshore disposal system using tankers and stations seems reasonable using established technologies.

(4) Clathrate formation may make it necessary to develop technologies using deep sea robots to maintain the smooth outflow of CO_2.

(5) In order to carry out macro evaluation of the influences on the ocean of residual CO_2, it is necessary to grasp the values of fundamental physical properties.

6.1 Remaining problems for evaluation

It is necessary to measure the physical values of CO_2 and to construct an exact model of its dynamic behaviour. The values which need to be measured include various kinds of physical constants, physical dissipation values, dissolution and the densities of CO_2 and its solutions, together with the problems of clathrate, including measurements of its formation speeds, conditions and stability. Field experiments concerning CO_2 diffusion may be necessary because of the various disturbances known to take place on the sea floor.

In view of social and environmental sensitivities it is important that there is a worldwide consensus before studies are commenced on this uncommon and unusual system. The existence of such a system cannot be justified by itself, without some prior discussion of a comparison of its economic and environmental merits and demerits with those of other means of handling the problem, including a comparison with methods of discharge into the atmosphere.

Figure 1
Basic concept of disposal systems to ocean

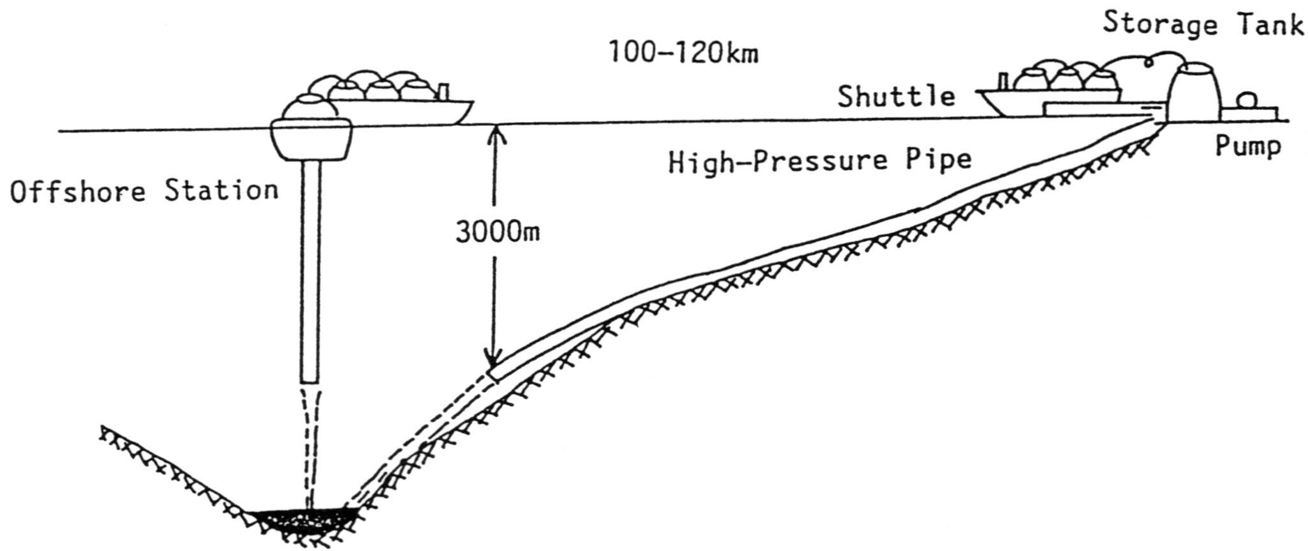

Table 1
Minimum energy for recovery, liquefaction and pumping of CO_2 to ocean

Items	Basic Equations	Energy (kJ/mol)
Recovery	$-RT\Sigma X_i \ell n X_i$	6.4
Liquefaction	$RT \ell n (P_2/P_1)$	15.3
Pumping	$(m_1 - m_2) gV$	0.04
Total		21.7

(CH_2) Thermal Energy 1 mol = 700 kJ
Efficiency to Elec. $\eta = 40\% = 280$ kJ
Combustion Eff. 21.7/700 = 0.031
Electric Efficiency 21.7/280 = 0.0775
Refrigerating Energy 24.5 kJ

Table 2
Function and element of the disposal systems

	Technical Element of each Function		
Functions	Solidification	Transportation(L.)	Pipeline (Liq.)
Recovery		○	
Solidification	○		
Liquefication		○	
Collection	Ref. Trans.	Hi.Pres. Ground Pipeline(Liquefied CO_2)	
Storage Tank	Ref. Tank	High Pressure Tank	
Transportation	Refrig. Ship	High Press. Tanker	Seabed Pipeline
Disposal	Sea Surface Disposal	Offshore Station Vertical Pipe	OverFlow from the end of Pipeline
Liq. Behaviour	Nat. Falling	Natural Dropping	Natural Flow

Table 3
Technical issues of each system element

System Element	Technical Issues	Feasibility	Note
Recovery	Imp. of Efficiency	f	
Liquefication	Imp. of Efficiency	f	
H.P. Pipeline	Space, Security	f	
H.P. TankYard	Site Selection	f	
H.P. Tanker	Operation, Reliabi.	f	
Seabed Pipeline	Laying, Maintenance	needs Develop.	
Vertical Pipe		f	
Offshore Station		f	
Pipe outlet	Beh. Anal.,	unknown	Clathrate
Natural Dropping	Beh. Anal., Opt. Depth	unknown	Flow shape, Cla
Natural Flow	Beh. Anal., Opt. Depth	unknown	Clathrate, Diff
CO_2 Pool Beh.	Behaviour of Diffus.	unknown	Diff. current

II.5 Observational Facts and Simulation of the Climate and Environment Changes in East Asia due to the Global Warming

R. Huang and H. Wang

Institute of Atmospheric Physics, Chinese Academy of Sciences, 100080 Zhong-Guen-Cun, Haiden-ku, Beijing, China

Abstract

This paper analyzes climate and environment changes in East Asia by using the observed data for the past 40 years. The analyzed results show that the decadal-scale climate change in East Asia during the past 40 years is very large, especially after 1965. There was a climate jump in climatic fluctuation of East Asia in about 1965. After 1965, precipitation has obviously decreased in north China and the Yellow river valley, but has increased in the Yangtze river valley and the Huaihe river valley in the 1980s. This has frequently caused drought disasters in north China and flood disasters in the Yangtze river valley and the Huaihe river valley. Moreover, the analyzed results also show that the dry trend in north China is analogous to that in the Sahel area of West Africa to some degree.

The frequent droughts and floods in north China, the Yangtze river valley and the Huaihe river valley have severely affected the grain yields and water resources in China.

The analyzed results also show that the decadal-scale climate change in East Asia is associated with the warming of the equatorial central and eastern Pacific and the global warming, due to the increase of CO_2. This increase of surface temperature and decrease of precipitation in north China have been simulated by using the climate model of IAP.

1 Introduction

It is well known that the problem of world climate change and its impacts on water resources and agriculture has become increasingly important, not only in the atmospheric sciences, but also in biological sciences.

As in other places in the world, a large climate change has occurred in East Asia and during the last 40 years has frequently caused climatic anomalies in north China, the Yangtze river valley and the Huaihe river valley (see Huang et al. 1992), bringing heavy droughts and floods to those places. These climatic disasters have a large impact on agriculture and water resources.

In order to explain the climate and environment changes in East Asia, and their impacts on the agriculture and water resources in China, a data set of precipitation and surface temperatures at 336 stations for 40 summers from 1950 to 1989 is used to look at the characteristics of climate change and the decadal-scale variation of climate in East Asia. The actual data of droughts and floods in various regions of China, and grain yields and water resources from 1950 to 1985 are analyzed. The possible cause of the climate change in East Asia is interpreted as the warming of the equatorial central and eastern Pacific and the global warming due to the increase of CO_2 in the atmosphere.

2 Climate Change and Decadal-Scale Variation of Climate in China During the Past 40 Years

2.1 Surface temperature

According to Bradley et al.'s (1984) investigation, shown in Figure 1, the general trend of surface temperature fluctuation (low frequency variation) in China has been similar to that in the Northern hemisphere as a whole over the last 105 years. The coldest period of the 20th century was in the early century from the 1900s to the 1920s, while a warming trend may be found from the 1920s to the 1940s, and the 1940s was the warmest period of the twentieth century. After the mid-1940s a cooling trend continued to the mid-1970s, followed by a warming trend. However, compared with the surface temperature fluctuation in the Northern hemisphere in detail, some differences may be found. These differences are that the cooling trend in China was more severe than that in the Northern hemisphere in the 1950s, but the warming trend was not much more than in the Northern hemisphere, and the temperature in the 1980s was much lower than that in the 1940s.

A warming trend was maintained in East Asia during the 1980s, but it was very different in different regions. Figure 2 shows the difference between the mean surface temperature in the 1980s and that in the 1950s (see Chen et al. 1991). From Figure 2, we can see that during the 1980s, a larger warming trend occurred in north China—in the northern part of north China, the winter mean surface temperature in the 1980s was about 1°C higher than in the 1950s. Generally speaking, the yearly mean surface temperature in the 1980s is about

0.44°C higher than that averaged for 1950-1979. However, the surface temperature in the Yangtze river valley and the Huaihe river valley decreased in the 1980s, ie a cooling trend occurred in the Yangtze river valley and the Huaihe river valley during the 1980s, which was contrary to the trend in north China. It may be found that during the 1980s, a warming trend also occurred in south China, although it was smaller than that in north China.

2.2 Precipitation

The change of precipitation in East Asia is more obvious than the change of surface temperature in the same region. In order to explain this climate change, a data set of precipitation at 336 stations in China is used to analyze the decadal-scale anomaly of precipitations in China.

Figure 3 shows the distributions of decadal mean anomaly percentage of summer precipitation in China in the 1950s (a), the 1960s (b), the 1970s (c) and the 1980s (d), respectively. As shown in Figure 3, there is a large climate change in the summer precipitation in China. In the 1950s, the summer precipitation was above normal both in the Yangtze river valley and the Huaihe river valley and in north and south China. This period may be called 'the Wet Period' in the climate fluctuation of China. However, in the 1960s, an obvious climate change occurred in north China. The summer precipitation abruptly dropped below normal in north China, the Yellow river valley, the Yangtze river valley and the Huaihe river valley. In the 1970s, this dry trend was continued in north China, the Yangtze river valley and the Huaihe river valley. Therefore, the 1970s may be called 'the Dry Period' in the climate fluctuation of China. Compared with the summer precipitation in the 1970s, in the 1980s a large climate change occurred in the Yangtze river valley and the Huaihe river valley. The summer precipitation greatly increased and was above normal there, but it decreased continuously in north and south China.

In order to see clearly this climate change, the 10 year running mean anomaly percentages of summer precipitation averaged for each region of China are made. Figures 4(a)-(d) are the 10 year running mean anomaly percentages of summer precipitation averaged for north China, the Yangtze river, the Huaihe river valley, the Yangtze river valley and south China, respectively. Figure 4 clearly shows the following results:

a) After 1965, the summer precipitation obviously decreased in north China and the Yellow river valley. This dry trend was severe in the 1980s.
b) In the 1980s, the summer precipitation obviously increased, and it was above normal in the Yangtze river valley and the Huaihe river valley, but decreased in south China.

From the results analysed above, it is found that a climate jump of precipitation occurred in East Asia in the mid-1960s, although a climate jump of temperature occurred in the 1950s, which was pointed out by Yamamoto et al. (1986). The characteristic of the climate jump is that the summer precipitation decreased abruptly and a dry trend occurred in north China. In order to see clearly the dry trend in north China, the 10 year running mean normalized anomaly of summer precipitation in north China is calculated (see Figure 5). As shown in Figure 5, the abrupt decrease of summer rainfall is very remarkable in north China after 1965.

The dry trend after 1965 in north China is associated with the global climate change, and it is analogous to the precipitation change in the Sahel area of West Africa. Figure 6 shows normalized anomalies of precipitation in the Sahel area (after the Fourth Annual Climate Assessment 1992). Figure 6 shows that during the 1950s, the precipitation was above normal, but precipitation became abruptly below normal after the mid-1960s and a severe dry trend occurred there. Comparing Figure 5 and Figure 6, it is clearly shown that there is a positive correlation between the precipitation anomalies in north China and those in the Sahel, which is in agreement with the investigation of Yan et al. (1990).

3 Impacts of the Climate Change in East Asia on Climatic Disaster, Grain Yields and Water Resources in China

3.1 Impacts on droughts and floods

The climate change in East Asia has brought many severe climatic disasters there, such as droughts and floods. It also has brought severe droughts to north China and the Yellow river valley, severe floods and droughts to the Yangtze river valley and the Huaihe river valley, and cooling summers in north-east China.

Table 1 shows the drought and flood summers in north China, the Yangtze river valley and the Huaihe river valley, respectively. It may be seen that after 1965, droughts, including general and severe droughts, frequently occurred in north China, and floods, including general and severe floods, mainly occurred in the Yangtze river valley and the Huaihe river valley and frequently occurred in the 1980s.

From Table 1 it may be also found that the frequency of climatic disasters including drought and flood disasters remarkably increased after 1965.

Recently, Huang and Li (1988) and Huang and Sun (1992) analysed the impacts of the western Pacific warm pool on the climatic anomalies in East Asia. The analysed results showed that the drought and flood summers in the Yangtze river valley and the Huaihe river valley are associated with the convective activities around the Philippines.

3.2 Impacts on grain yields

Climatic disasters including droughts, floods and cooling summers heavily affect the growth of crops. Thus, the frequent climatic disasters have brought a decrease of grain yields.

According to the statistics, the averaged decrease of grain yields due to climatic and synoptic disasters in China is more than 10 billion kg each year. As Figure 7 shows, the largest percentage of decreased grain yields is due to drought disaster, and it is about 50% of the total decrease resulting from all climatic and synoptic disasters. The decreased grain yield due to flood disasters is about 27.6% of the total decreased yield. Thus, in China the economic loss due to droughts is larger than all other natural disasters.

The decrease of the total yields of grain production due to climatic disasters, including droughts and floods, for 1951-1980 in various regions of China is analysed. The proportion of the decreased grain yield due to climatic disasters is larger in north China, the Yangtze river valley and the Huaihe river valley. The proportion of the decreased grain yield in these two regions is more than 50% of the total decreased grain yield due to all climatic and synoptic disasters in the whole of China. This may be a result of the frequent climatic disasters occurring in these two regions.

3.3 Impacts on water resources

In the previous subsection, it was mentioned that there is a dry trend in the climatic fluctuation of north China after 1965, especially in the period from the late-1970s to the early-1990s, and droughts frequently occurred there. This prolonged drought has caused the decrease of water resources in north China. Figures 8 and 9 are decadal-scale variations of the anomaly percentage of river run off and water resources in north China. From these two figures, it may be seen that the river run off and water resources have successively decreased from the 1970s, especially in the 1980s. In the 1980s, the river run off was below normal and has decreased about 50% from that in the 1950s. Similarly, the water resources were below normal in the 1980s and have decreased about 60% from that in the 1950s. The lack of water may become an increasingly severe problem for the development of industry and agriculture in north China and in the Yellow river valley.

3.4 Possible cause of the climate change in East Asia

According to the simulation of global warming by climate model, the largest warming area should be in the high latitudes. However, Nitta's (1990) analysis from the observed data showed that the largest warming area of ocean is in the equatorial central and eastern Pacific. In order to investigate the cause of the climate change in East Asia, the variation of SST in the equatorial central and eastern Pacific is analysed. Figures 10 and 11 are the five year running mean anomalies of SST in the equatorial central and eastern Pacific, respectively. From Figures 10(a) and 10(b), it may be found that the SST in the equatorial central Pacific in the winter of the 1980s is about 1°C higher than that in the winters of the 1970s. Similarly, as shown in Figures 11(a) and 11(b), the SST in the equatorial eastern Pacific in the winters and summers of the 1980s are about 0.8°C and 0.5°C higher than in the winters and summers of the 1970s, respectively. Therefore it is very obvious that the SST in the equatorial central and eastern Pacific in the 1980s is about 0.8°C higher than that in the 1970s. This may indicate that the equatorial central and eastern Pacific is warming remarkably.

This 'decadal-scale ENSO event' appeared to occur in the 1980s, while a 'decadal-scale La Nina' occurred in the 1970s. Thus, there may be a 'decadal-scale ENSO cycle' in the anomaly variations of SST in the equatorial central and eastern Pacific.

Huang and Wu's (1989) investigation showed that the ENSO events heavily influence precipitation in the eastern part of China. As shown in Figure 12(a), in the developing stage of an ENSO event, the SST in the equatorial central and eastern Pacific is higher, ie the equatorial eastern Pacific is in a warm episode, the summer monsoon rainfall is strong in the Yangtze river valley and the Huaihe river valley of China, but it is weak in north and south China. Thus, in the warm episode of equatorial eastern Pacific, floods used to occur in the Yangtze river valley and the Huaihe river valley, while droughts used to occur in north and south China. On the contrary, as shown in Figure 12(b), in the decaying stage of the ENSO event, the SST in the equatorial central and eastern Pacific is lower, ie the equatorial eastern Pacific is in a cold episode, the summer monsoon rainfall is weak in the Yangtze river valley and the Huaihe river valley, but it is strong in north and south China. Thus, in the cold episode of equatorial eastern Pacific, droughts used to occur in the Yangtze river valley and the Huaihe river valley, while floods used to occur in south China and the summer rainfall used to be normal or above normal in north China.

From the analyses mentioned above, the 'decadal-scale ENSO event' which occurred in the 1980s may cause the increase of summer rainfall in the Yangtze river valley and the Huaihe river valley and the decrease of summer rainfall in north and south China. This may partly be interpreted that in the 1980s, floods increased in the Yangtze river valley and the Huaihe river valley, while droughts frequently occurred in north China.

5 Numerical Simulation of the Climate Change Caused by CO_2 Doubling

The possible cause of the climate change in East Asia is an interesting research problem. Most meteorologists believe that the impact of human actions is one of the most important causes of climate change above natural variability. The approach to evaluating the impact of increased greenhouse gases due to human actions may be the numerical simulation by climate models using super computers. It has been more than a decade since meteorologists began to study the CO_2-induced climate change by climate models, but many uncertainties still exist in judging the warming magnitude and regional distribution of climate change etc. Figure 13 shows the CO_2 doubling-induced change of soil moisture in East Asia in summer simulated by NCAR, GISS, GFDL, UKMO and OSU climate models (Zhao, et al., 1988). Obviously, the change of soil moisture in East Asia induced by CO_2 doubling is very different with the using climate model. For example, the CO_2 doubling-induced change of soil moisture by the NCAR model is a dry trend in north-east China, north China, south China and the India-

China Peninsula and a wet trend in east China, the Korean Peninsula, Japan and north-west China. This result is in some agreement with the result simulated by the GISS model although these two models obtain opposite results in south-west China and India. However, the CO_2 doubling-induced change of soil moisture by GFDL is a dry trend in North Asia, East Asia and Central Asia, and a wet trend in South Asia. The simulated result by the UKMO model is in part similar to that induced by the GFDL model although a wet trend can be found in south China in the results of the UKMO model. The simulated result from the OSU model is a wet trend in east China, south China, the India-China Peninsula and India. Thus, there are many significant differences between the distributions of CO_2 doubling-induced change of soil moisture by various climate models. Therefore, it still is uncertain what climate change occurs in East Asia when CO_2 increases in the atmosphere.

In order to investigate the climate change induced by CO_2 doubling, the two level IAP-AGCM coupled to a 60m deep mixed layer ocean model and a zero-layer thermodynamical model is used to simulate the climatic effects of CO_2 doubling in the atmosphere. The simulated results are as follows:

5.1 Surface temperature

Figure 14 is the simulated global distribution of surface temperature change induced by CO_2 doubling. We can find that when CO_2 increases in the atmosphere, a warming trend in the northern Hemisphere is larger than that in the southern Hemisphere, the largest warming regions are in the Bering strait, the north-east part of North America and Greenland, respectively. Moreover, it can be seen that the warming magnitude in the Tibetan Plateau is also very large. In order to see the climate change in East Asia induced by CO_2 doubling in detail, a simulated distribution of surface temperature change in China caused by CO_2 doubling is given in Figure 15. It may be clearly seen from Figure 15 that when CO_2 increases in the atmosphere, the surface temperature increases obviously in China, especially in the Tibetan Plateau and south China. Compared to the observational facts, this simulated warming trend is in agreement with the observational result, except in the Yangtze river valley and the Huaihe river valley the observation appears to show a cooling trend.

5.2 Precipitation

Figure 16 shows the simulated global distribution of precipitation change induced by CO_2 doubling. Obviously, when CO_2 increases in the atmosphere, a dry trend will occur in the subtropics and a wet trend may occur in the high latitudes of the northern Hemisphere. In order to see the precipitation change in East Asia induced by CO_2 doubling in detail, a simulated distribution of precipitation change in China in summer caused by CO_2 doubling is given in Figure 17. From Figure 17 we find that when CO_2 increases in the atmosphere, a dry trend will appear in north-east China, the Yellow river valley and south China and a wet trend may appear in the central and lower reaches of the Yangtze river in summer. Compared to the observational fact, this simulated result is in some agreement with the observational precipitation change.

Therefore, the results mentioned above explain that the 'greenhouse effect' may play an important role in the dry trend in north China and the Yellow river valley.

6 Summary

Climate and environment changes in East Asia are analyzed by using the observed data for the past 40 years. The results show that the decadel-scale climate changes in East Asia during the past 40 years are very large, especially after 1965. There was a climate jump in the climatic fluctuation in East Asia in about 1965. After 1965, precipitation has obviously decreased in north China and the Yellow river valley, while the surface temperature has increased. Thus, the dry trend is increasingly severe in north China. Moreover, the analyzed results also show that the dry trend in north China is analogous to that in the Sahel area of West Africa in some degree.

On the contrary to the climate change in north China, the precipitation has obviously increased in the Yangtze river valley and the Huaihe river valley from the late 1970s, where flood disasters frequently occurred.

The frequent droughts and floods in north China and the Yangtze river valley and the Huaihe river valley have greatly influenced the grain yields and the water resources in China. Among them, the economical loss due to drought is larger than the total losses caused by all natural disasters.

The analyses also show that the climate and environment changes in East Asia may be associated with the warming of the equatorial central and eastern Pacific and the global warming due to the greenhouse effect. Moreover, the two level IAP-AGCM coupled to a 60m deep mixed layer ocean model and a zero-layer thermodynamical model is used to simulate the effects of CO_2 doubling in the atmosphere on the climate and environment changes in East Asia. The simulated result shows that when CO_2 increases in the atmosphere, the surface temperature increases and the precipitation decreases in north China. This result is in agreement with the recent climate change in north China.

According to the recent investigations and this study, it is very difficult to distinguish the natural climate variability from the global climate change. Thus, the natural climate variability and its implication in global warming should be an important factor in studies on climate and environment.

References

Bradley, R.S., H.F. Diaz, P.D. Jones and P.M. Kelley, 1984: 'Secular fluctuations of temperature over northern Hemisphere land', *The Climate of China and Global Climate*, Proceedings of the Beijing International Symposium on Climate. 30 Oct – 3 Nov 1984, Beijing, China, 76–87.

Chen, L.X., Y.N. Shao, M. Dong, Z.H. Ren and G.S. Tian, 1991: 'Preliminary analysis of climate variation during the last 39 years in China'. Adv. Atmos. Sci., Vol. 8, 279–288.

Huang, R.H. and W.J. Li, 1988: 'Influence of the heat source anomaly over the western tropical Pacific on the subtropical high in summer and its mechanism'. Scientia Atmospherica Sinica (special issue), 107–116.

Huang, R.H. and Y.F. Wu, 1989: 'The influence of ENSO on the summer climate change in China and its mechanism'. Adv. Atmos. Sci., Vol. 6, 21–33.

Huang, R.H., Y.L. Liang and L.C. Song, 1992: 'Variability of summer droughts and floods in China during the recent 40 years and primary investigation of its causes, Investigation on Some Problems in Climate Change', LASG Monography, No. 2, 14–29.

Huang, R.H. and F.Y. Sun, 1992: 'Impacts of the tropical western Pacific on the East Asian summer monsoon'. J. Meteor. Soc. Japan, Vol. 70, No. 1B, 243–256.

Nitta, Ts., and S. Yamata, 1989: 'Recent warming of tropical sea surface temperature and its relationship to the northern Hemisphere circulation'. J. Meteor. Soc. Japan, Vol. 67, 375–383.

Wang, H.J., Q.C. Zeng and X.H. Zhang, 1993: 'The numerical simulation of the climate change caused by CO_2 doubling'. Science in China (Series B), Vol. 36, 451–462.

Yamamoto, R., T. Iwashima, N.K. Sanga and M. Hoshihai, 1986: 'An analysis of climatic jump'. J. Meteor. Soc. Japan, Vol. 64, 273–281.

Yan, Z.H., J.J. Ji and D.E. Ye, 1990: 'Northern Hemispheric summer climatic jump in 1960s'. Science in China (Series B), Vol. 33, 1092–1101.

Zhao, Z.C. and W.W. Kellogg, 1988: 'Sensitivity of soil moisture to doubling of Carbon Dioxide in climate model experiments, Part II: The Asian Monsoon Region'. J. Climate, Vol. 1, 367–377.

Figure 1
The low-frequency fluctuation of five year running mean surface temperature of China (thin line) and the Northern hemisphere (thick line). The temperature averaged for 1880-1884 has been chosen as the zero reference point of the Northern hemisphere.

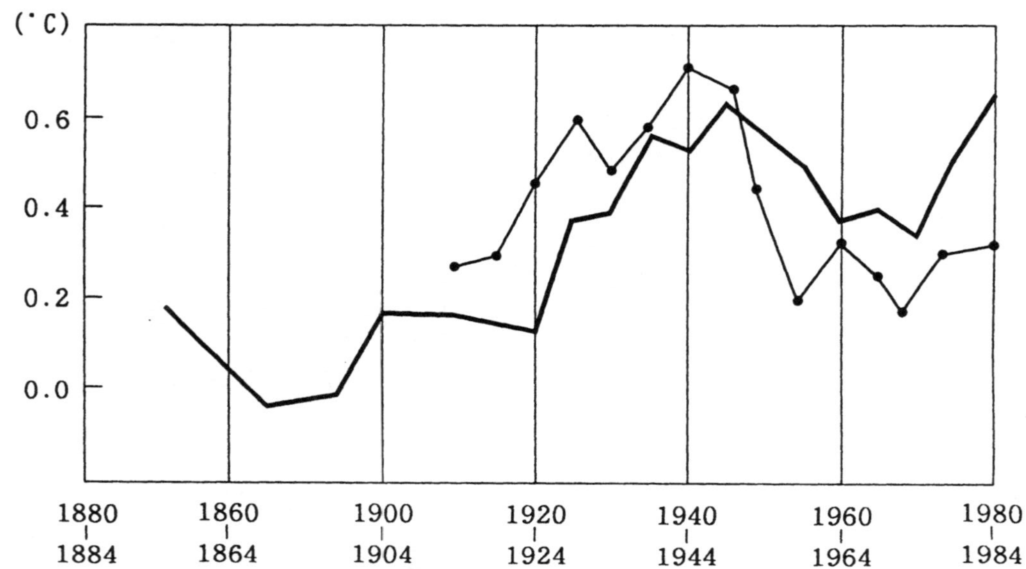

Figure 2

The distributions of winter (a) and summer (b) mean temperature (°C) differences between the 1980s and the 1950s in China. The cooling regions are shaded and interval of contours is 0.5°C (after Chen et al. 1991).

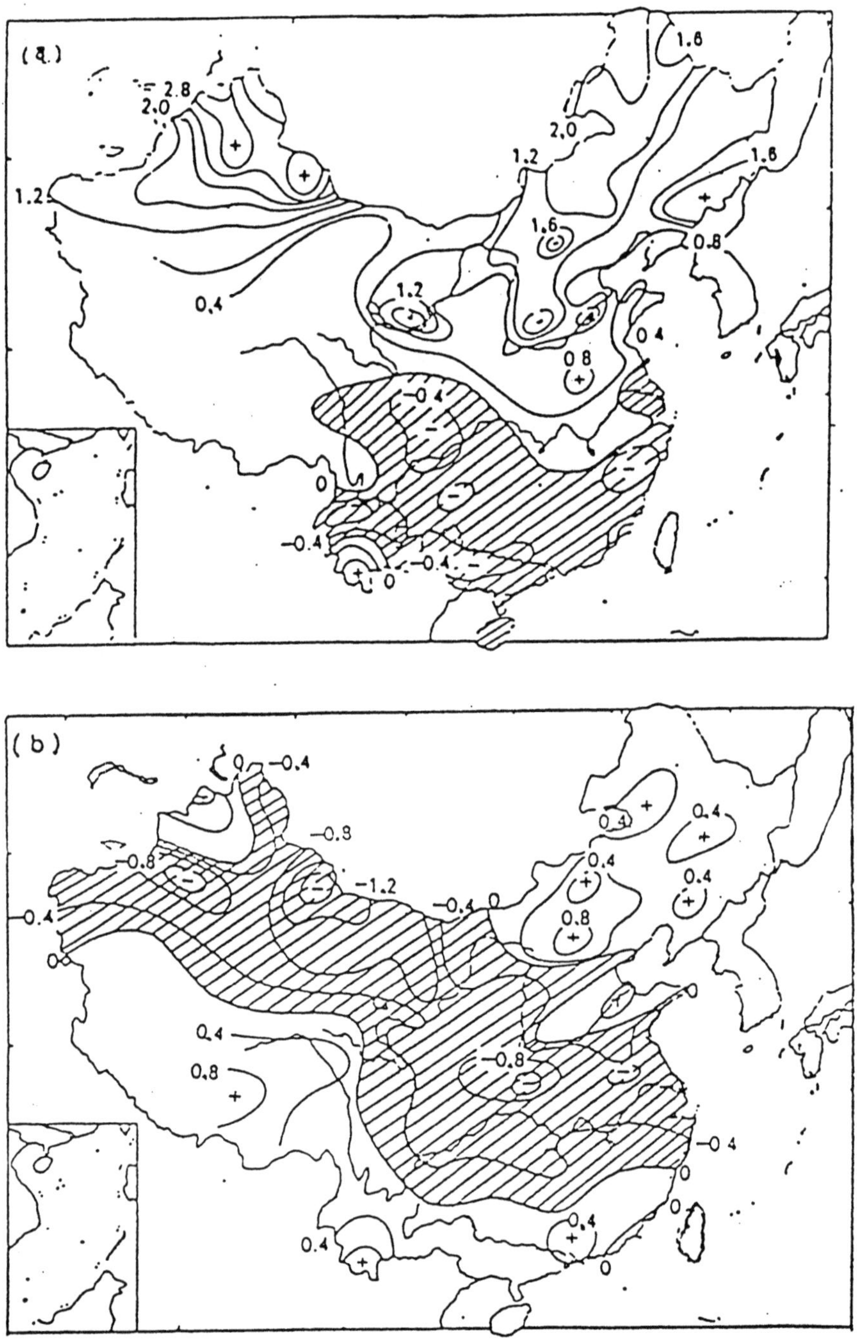

Figure 3
The distributions of summer mean anomaly percentages (%) of precipitation in China in the 1950s (a), the 1960s (b), the 1970s (c) and the 1980s (d).

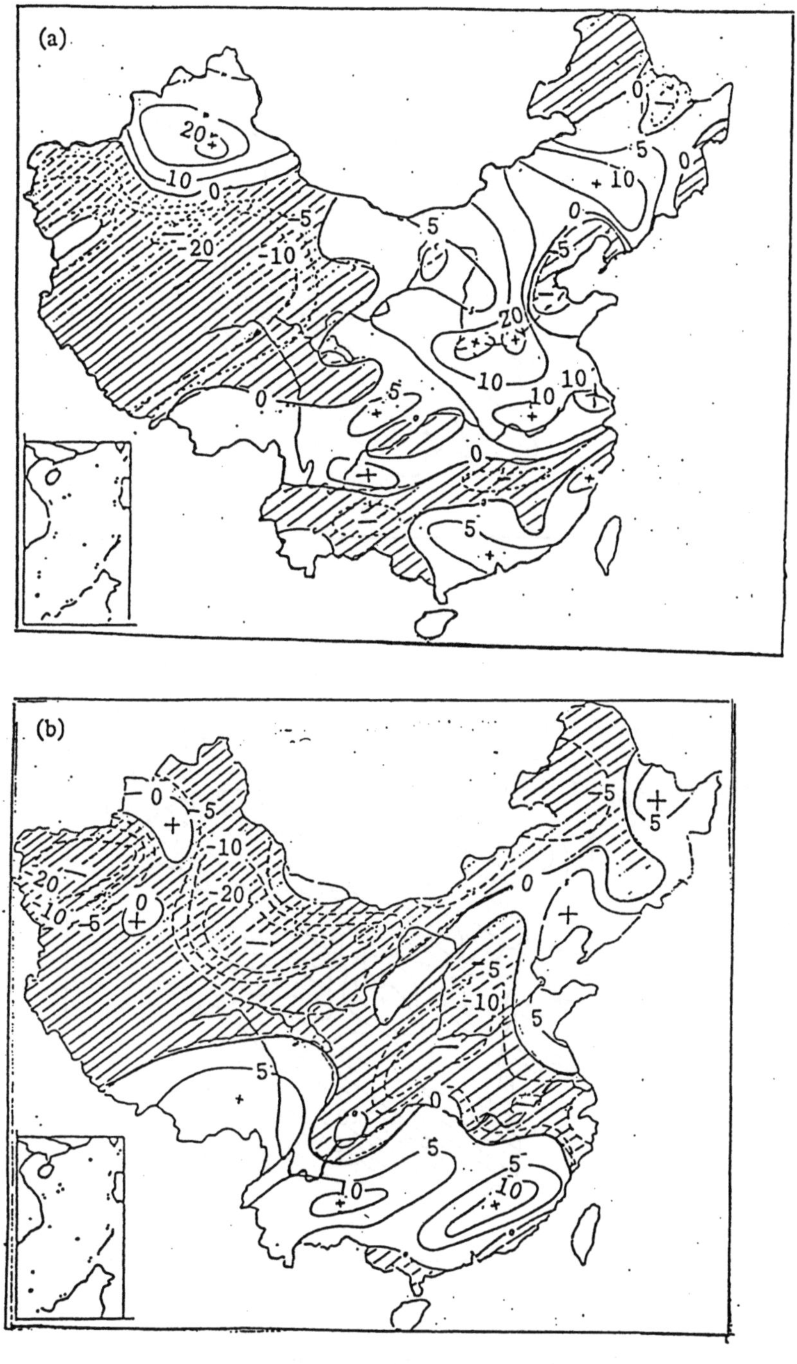

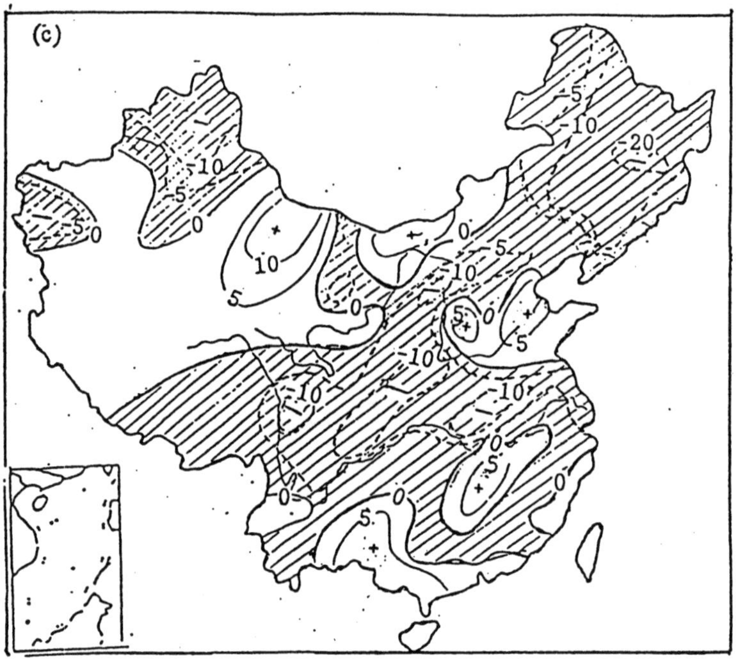
(c)

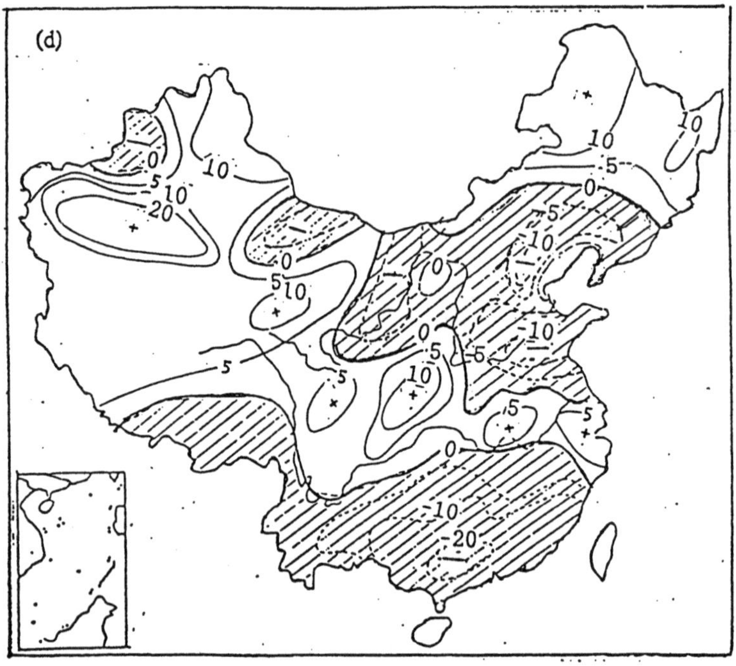
(d)

Figure 4

Ten year running mean anomaly percentages (%) of summer precipitation averaged for north China (a), the Yangtze river and the Huaihe river valley (b), the Yangtze river valley (c) and south China (d).

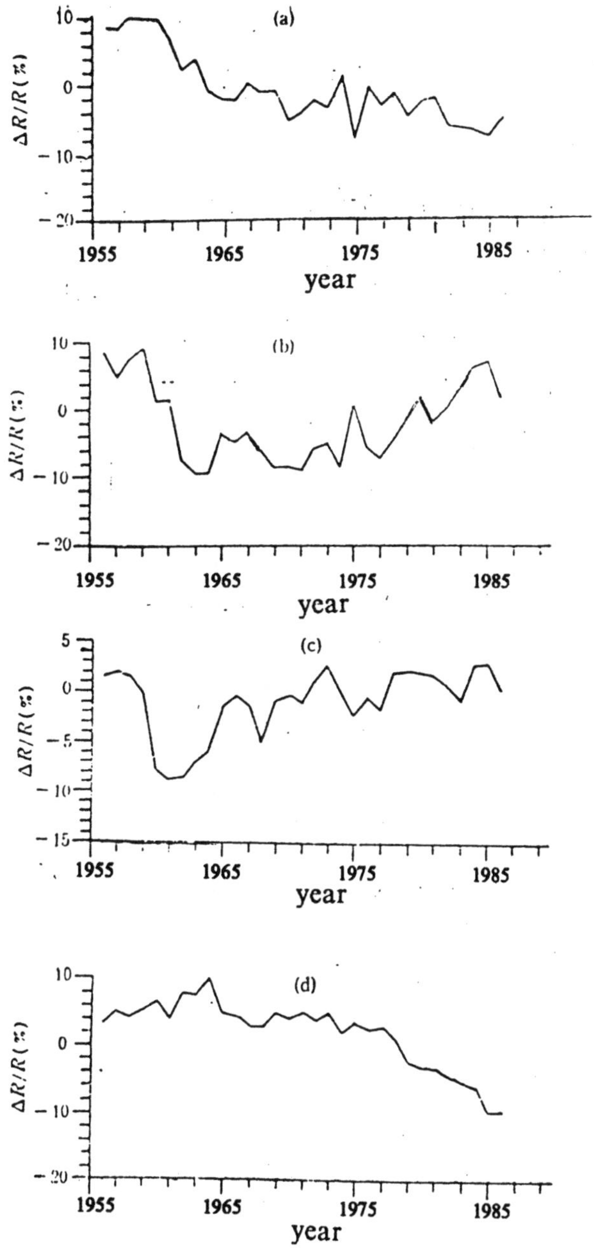

Figure 5
Ten year running mean normalized anomalies of summer precipitation in north China.

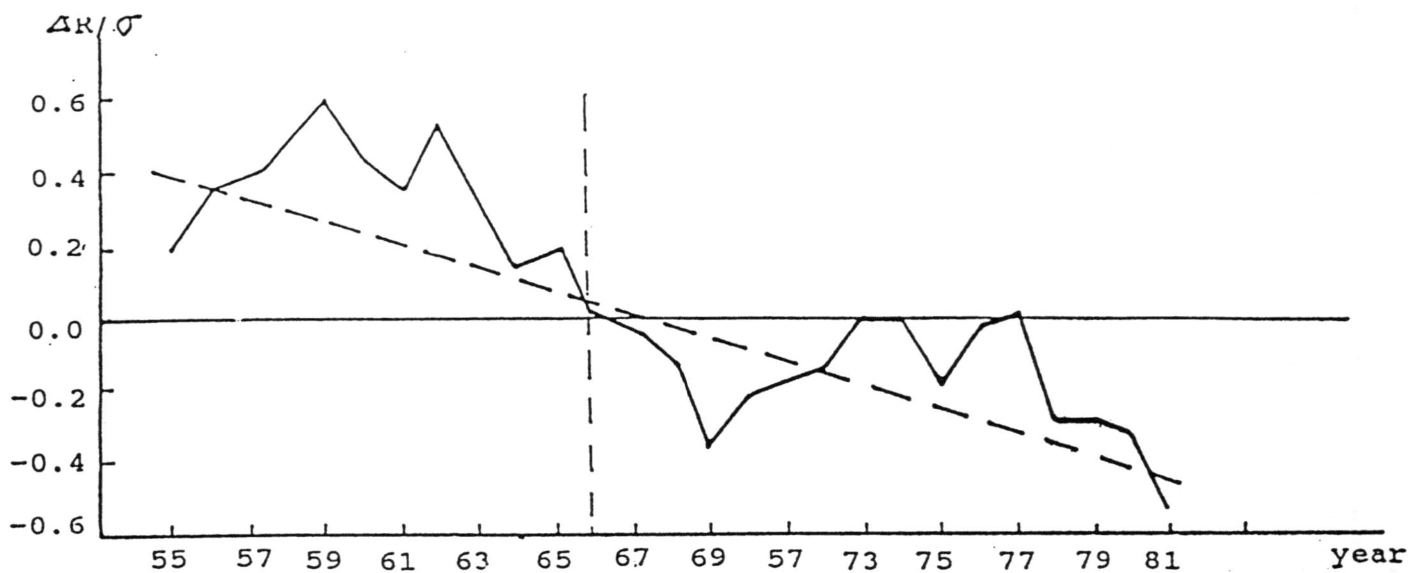

Figure 6
Normalized precipitation anomalies in the Sahel area. (after Fourth Annual Climate Assessment 1992)

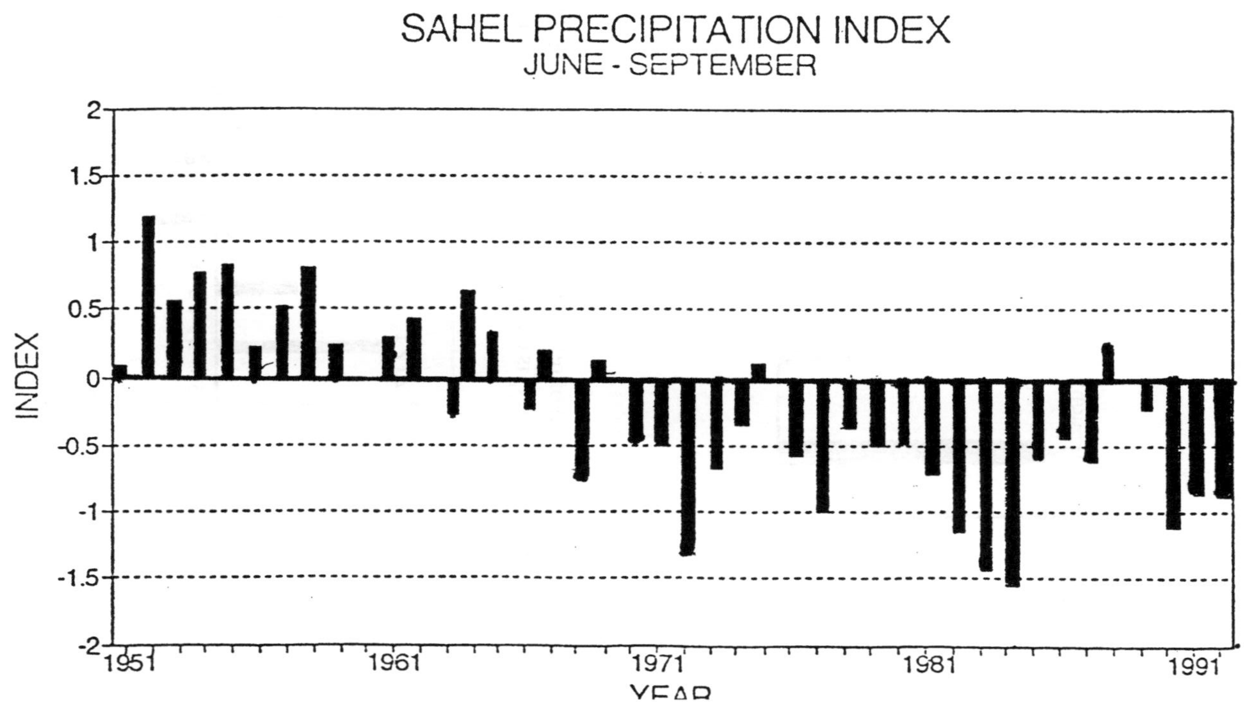

Figure 7
The proportion of grain yields decrease due to a climatic disaster to total decrease of grain yields due to all climatic disasters.

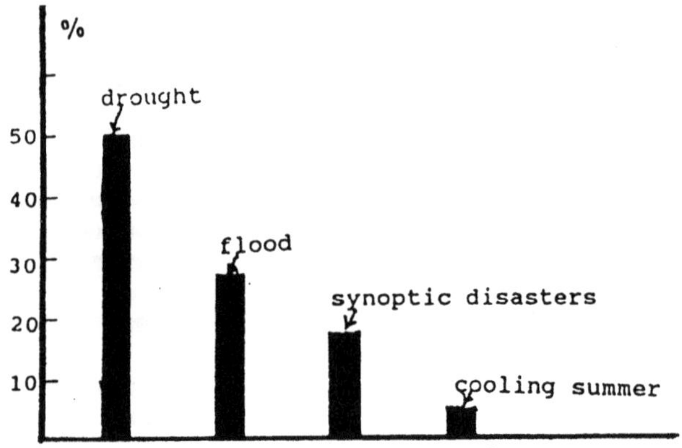

Figure 8
Decadal-scale variation of relative anomalies of river run off in north China. r indicates anomalies and r_0 is climatological mean of river run off.

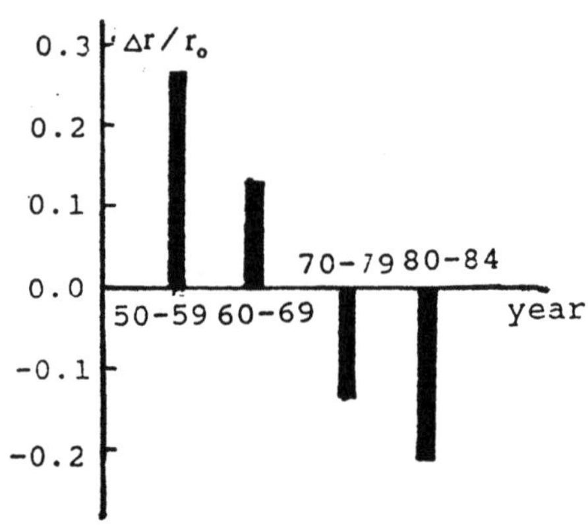

Figure 9
Decadal-scale variation of relative anomalies of water resources in north China. w indicates anomalies and w_0 is climatological mean of water resources.

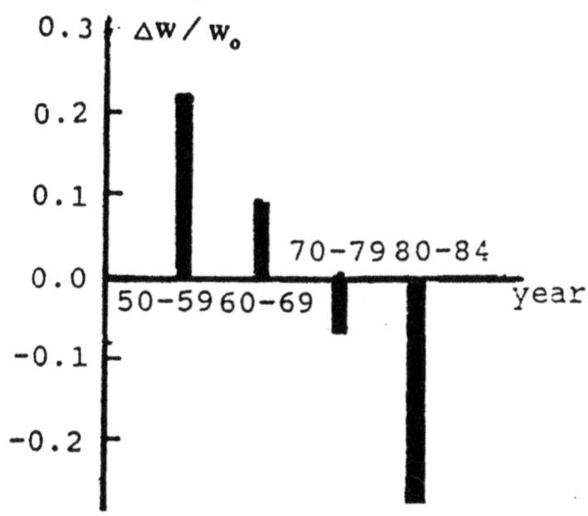

Figure 10

Five year running mean SST anomalies in the equatorial eastern Pacific in winter (a) and summer (b).

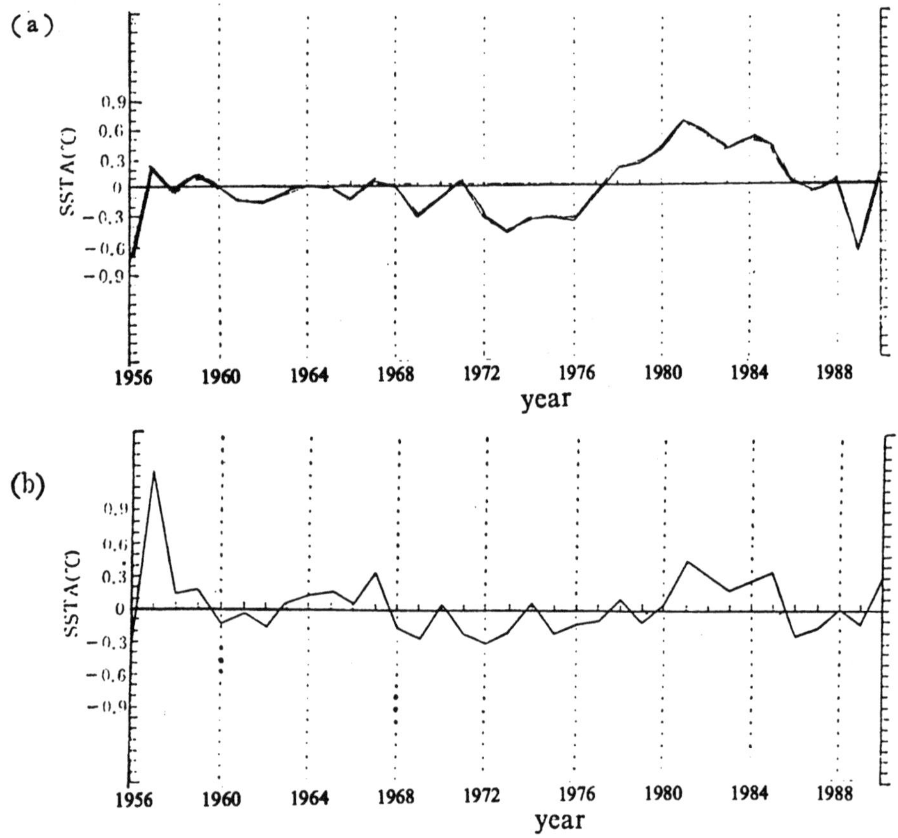

Figure 11

Five year running mean SST anomalies in the equatorial central Pacific in winter (broken line) and summer (solid line).

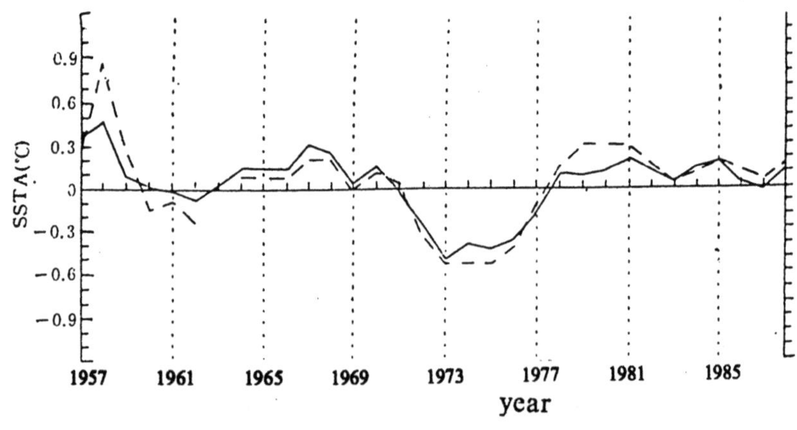

Figure 12

The correlation between the summer rainfall in China and the SST in the equatorial eastern Pacific in the developing (a) and decaying (b) stages of ENSO event. Shaded areas indicate the correlation coefficients above 0.4.

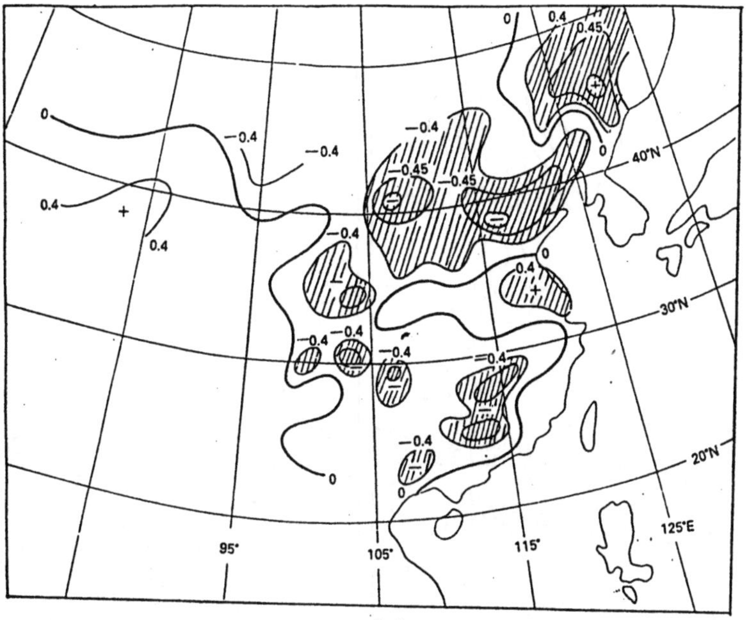

(a)

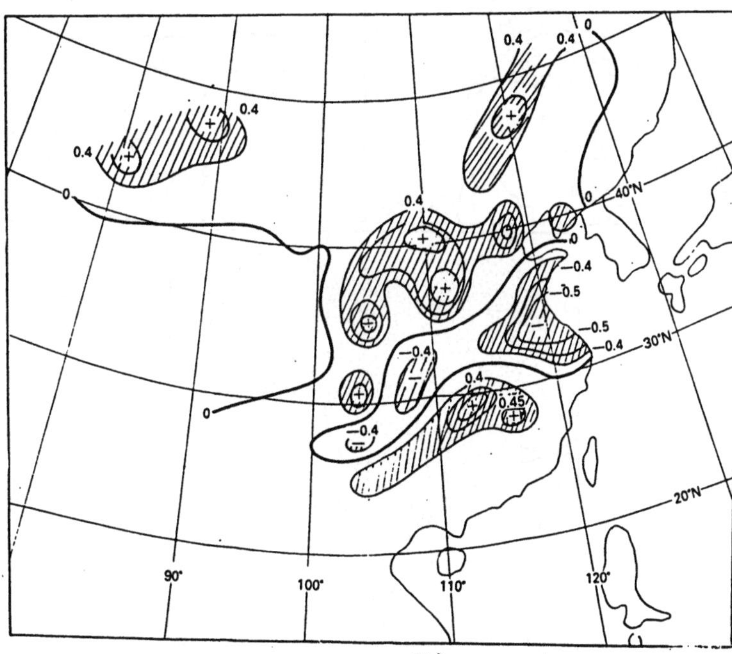

(b)

Figure 13

The distributions of CO_2 doubling-induced soil moisture change simulated by the climate models of NCAR, GISS, GFDL, UKMO and OSU. The shaded areas are those where there is a decrease or change to a drier condition (after Zhao et al., 1988). Units are cm of water.

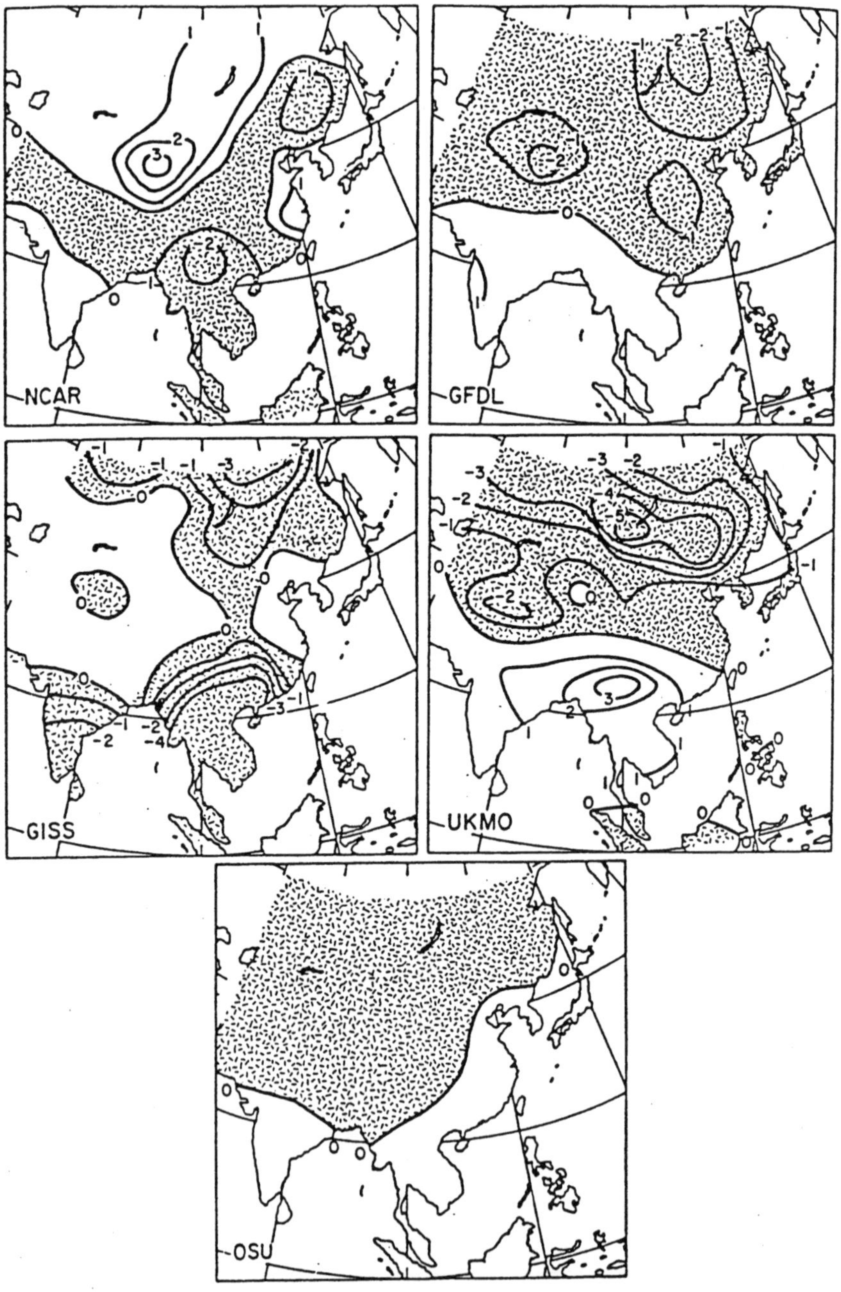

Figure 14
The global distribution of CO_2 doubling-induced surface temperature change simulated by using the climate model of IAP. Units are °C.

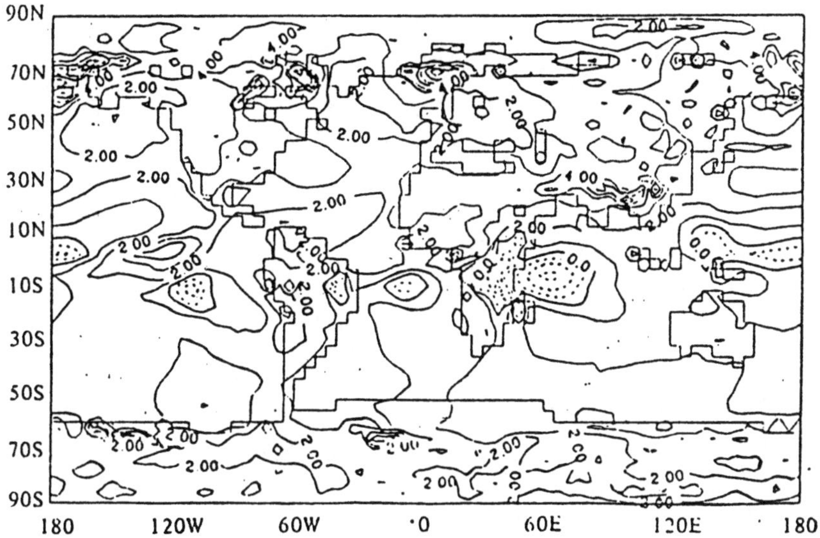

Figure 15
Same as Figure 14, but for the distribution in China.

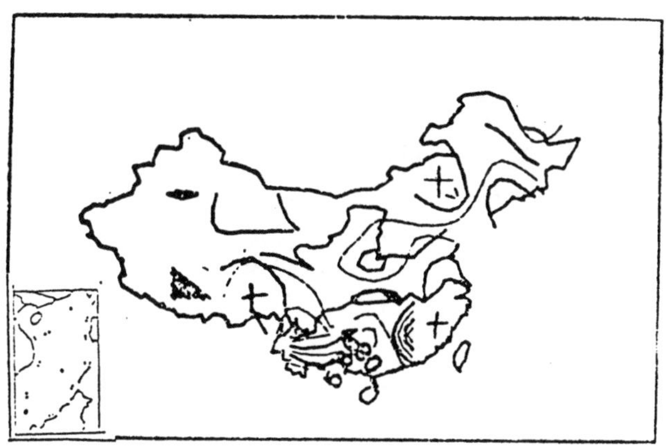

Figure 16

The global distribution of CO_2 doubling-induced summer precipitation simulated by using the climate model of IAP. Units are mm/day.

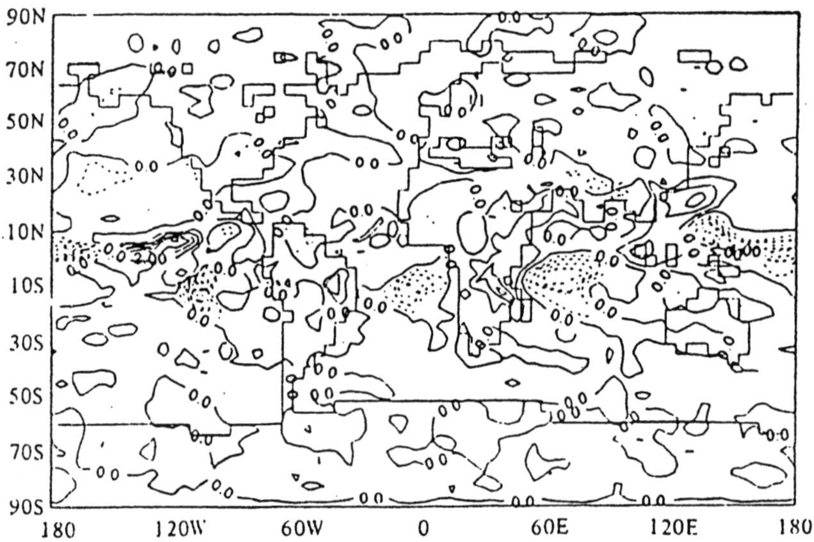

Figure 17

Same as Figure 16 but for the summer precipitation in China.

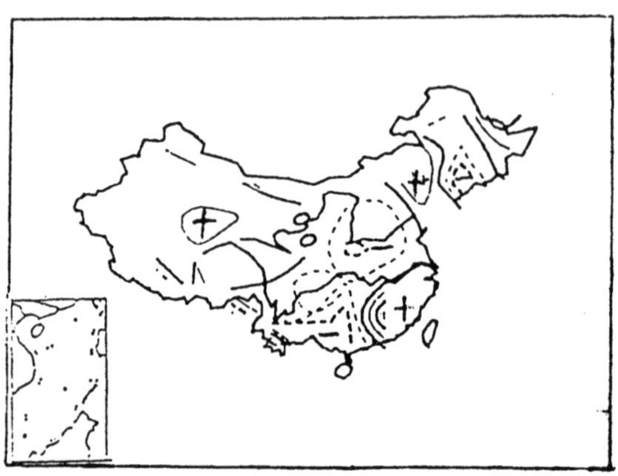

II.6 Remarks

Professor Mori

Here I would like to talk about recyclable commodities as well as natural recycling for the purpose of efficient energy use. The key factor in the new energy recycling resources is to substitute as yet untouched sources with newly processed materials or efficient energy resources. We already possess the technological possibilities to improve on. These technologies yet need to be supported by social systems regarding how to utilize reprocessed energy and materials.

What then could be the bottlenecks of the system? The bottlenecks do exist in three stages. First, how to develop the technologies and the resource and capital allocation. Secondly, how the technologies should be implemented. Economic recession, infrastructure regression, and public inacceptance could be a bottleneck for the implementation. Then thirdly, how the technologies should be diffused all over the world. The interdisciplinary cooperation between governments, companies and researchers—and then between national and international corporations—will all be needed to make the breakthroughs. From this point of view, what is important is not only to develop technologies based on engineering but also to assess what could be the barriers in the social system.

Professor Sudara

People take it that recycling is a way to cope with environmental problems, but actually it is not. Because recycling brings something back—reprocesses it again—we are using more and more energy and chemicals. It doesn't end up by cleaning up everything. I would argue that rather than recycling we should push on conservation of resources and of energy. Take paper, for example. You cannot recycle it indefinitely. You can recycle it only up to a maximum of seven times, but usually it is three times and you have to add a lot of chemicals, some very toxic.

Secondly, environmental problems should not be handed over from the North to the South. Rather, they should be tackled by mutual initiatives and cooperation among all the countries concerned. The developed countries have always tried to tell the developing countries that if you want to make a correction of your pollution problems, come to us, we have something to sell to you. But actually some chemicals have been shipped over from the developed countries. I think this is one of the problems we have to take into consideration.

In addition, we are getting more and more into asking for more energy. There is no definite plan as to how much a country should use energy. The time will come when we will see the upper limits of the available resources and population increase. Then each country will have to solve its energy problems, rather than try to expand further. There are a lot of techniques that can be used for energy conservation. The developing countries, however, always look to the developed countries and follow in their footsteps. I am really impressed with the way in which the photosynthesis system is expanding in this country. Most of the Southern countries are around the area where lots of solar energy is available. This is one of the lines which we should try to help each other with. The superconductor can also be initiated in many of the developing countries. In Thailand, for example, we are trying to work on that. If there is more cooperation this can go faster.

I am very scared to see that you are again thinking of dumping carbon dioxide waste into the ocean. The ocean is a world system. If you dump something without careful consideration, then finally it will come back and it will really hurt the whole system. The global system is such that everything is interconnected. Maybe carbon dioxide can be used for the process of photosynthesis, rather than sinking it down to the deep water. There are a lot of studies right now on the sinking of carbon dioxide in the ocean.

On global warming, I think the essential part has not been touched upon. When you talk about the shifting of the rim fall and the temperature, you have to talk about how to change the use of land as well. The use of the land can greatly affect atmospheric conditions, particularly when the forest is cleared or when it is replanted. The amount of rainfall will be different. The forest itself absorbs moisture from the atmosphere and then releases it again through the active process of transpiration. We should also be very careful not to jump to the conclusion that this is really what is happening, that it is getting warmer and warmer. And I think we also have to look at how the environmental condition is affecting the people who are growing crops in an area as well.

I would like to end my discussion by saying that we have to try to work together, not only with modern technology itself but also by looking back into the traditional ways of using things and try to find out how we can get the best use of our resources.

Professor Murota

I would like to make some remarks mainly upon Dr Levine's presentation. The first relates to the possibilities of an explosion in energy demand in developing countries. If they continue with rapid development through heavy industrialization, then we may have what we call 'energy demand explosion'. We have three examples here. In the 1920s the United States had a peak and in the 1930s the U.K. did the same. In the 1960s it was Japan. A problem arises when the growth of energy demand is higher than the economic growth rate. This brings me to a question on China. Dr Levine argues that the growth rate of the GDP is far higher than energy demand and that this is mainly because of the success in their conservation policies. But I would like to show that there are two other possibilities. The first is that energy source may be shifted from coal to oil. Oil is more expensive than coal. This was what happened in Japan and other developed countries. When this kind of transformation of energy sources happens, then the growth rate of the energy demand may decrease compared with that of the GDP. The second possibility is that because China is changing from a socialist system to a market system they are experiencing a growth in material production. In that case, the reduction in energy demand may be underestimated. When productivity increase in the private sector is lower than the industrial output this may cause higher GDP.

On the question of sustainability . . . we are now conducting a long economic forecast looking at each country's and each region's economic and population growth up to the year 2000. Currently we have a greater degree of income inequality between North and South. Even if we have optimistic assumptions about the future of the developing countries, still for the coming 20 or 30 years this kind of economic inequality will not be improved. As you know, poverty is the largest environmental problem in developing countries. As long as the present system is based upon our current technology structure, I am a bit skeptical about its sustainability, and about the technologies we already have in place.

II.7 Session Discussions

Dr Levine

First, there will be phases through which countries will go and in those phases the energy intensity may in fact increase for a time and then later decline. But that makes the situation somewhat more worrisome, because some of these forecasts for demand growth are factors which may well push it up even more. That doesn't change the importance of energy efficiency in any way. In fact, in some ways you have to push even harder on energy efficiency if you have other things pushing you toward energy supply.

On China, on the other hand, I do not agree. For China, during the period of greatest energy efficiency, the fraction of energy from coal remained steady at more than 70 percent and the fraction from oil remained steady at somewhere around 25 percent. Thus it wasn't a switch in fuel form. That had occurred earlier. In fact, the Chinese suggest that only about half of the energy intensity decline is due to energy efficiency—maybe even less than that. Our analysis suggests that about 80 percent of that decline in China is actually a result of energy efficiency. Indeed, there are lots of accounting problems.

Professor Huang

I have a comment to Dr Ishitani. From the viewpoint of global environment protection in the future, it is not possible to fix CO_2 into the deeper water. According to the dynamic structure of the sea currents and the stratification of ocean temperature, CO_2 cannot transport from the deep ocean into upper water. But in the future, due to the climate changes, the oceanic state may change. In the long term, CO_2 disposal into the deeper ocean may be possible.

Professor Ishitani

All we can do for the moment is conduct a feasibility study of a technological possibility. We don't think this is very desirable, but still we are thinking about whether it is feasible or not. If we want to use coal for more than 300 years, we should sequester the CO_2 somewhere even though the greenhouse effect is unavoidable. In such a case, it is quite natural that we must search for a place where we can insulate CO_2 from the atmosphere. On the dynamics of the ocean, we are also very much interested in mathematical models of the global dynamic current. Max Plank Institute is most advanced in this respect. According to their research, once the CO_2 is disposed into the ocean it will come up to the atmosphere again after 1000 years. In 1000 years, probably we can develop a more efficient and more rational system.

Dr Salim

First, Mr Ishitani mentioned about the practical feasibility of dumping CO_2 in the ocean. The problem is economic feasibility, however. We are polluting the air because we assume that it is a free resource. Is the sea also a free resource? Could it be economically feasible? Mr Ishitani's study argues on the other hand that there is a need for global cooperation for the protection of the sea resource. If everybody does CO_2 dumping in the sea, then we end up with the whole sea of the globe as a dumping ground. On the point raised by Mr Huang, the climate change has been assessed because of CO_2. What is not clear then is where does the CO_2 come from? Is it because of the pollution by Japan?

Professor Yamada

Dr Sudara spoke about the importance of photosynthesis in the ocean. Perhaps by increasing the photosynthesis rate in the ocean, the CO_2 also falls into the deep sea. There is no difference between the dumping of CO_2 into the deep ocean and the fixing of it by photosynthesis. Of course the density of the CO_2 in the ocean is very different. Depending on the CO_2 dynamics in the ocean, perhaps we can select either the photosynthetic promotion or the CO_2 dumping. I don't like the dumping of CO_2 into the ocean and also I don't like to decrease the energy intensity to sequester CO_2 into the deep ocean. A technical feasibility study of this is very important in order to understand the ocean's relationship to the CO_2.

Professor Huang

In our numerical model, the amount of CO_2 has increased considerably. In order to maintain the eco-system, it is important not to just dump it, but to increase the phytoplankton which can use the CO_2 in the ocean. In this way you can begin to increase the production of food to match the increase of population on land. But we have to be very careful about how much we should increase it. If it is too much, then it will become a red tide and when it decays it will cause other problems.

Professor Yamada

Dr Salim pointed out that we didn't evaluate environmental cost under water for the moment. One of the reasons is that at least CO_2 is not poisonous like nuclear waste in the sea. According to some oceanologists, however, if we can place CO_2 adequately, then probably the production rate of creatures in the ocean will be increased. But the problem is, of course, the underwater creatures. It is quite difficult to estimate the cost of killing such creatures. For the moment, it would be more appropriate to disperse the CO_2 in the sea water below 500 metres. In that case, the CO_2 will come up to the atmosphere in 1000 years.

Dr Nakamura

In my view just one thing is lacking, and that is the importance of pollution control at the same time when energy efficiency is being increased. It may be of interest to recall that Japanese anti-pollution activities began in the early 1960s, first with an attack on particulates and sulphoxides. Then, such programs were adversely affected by activities against noxious emissions. These anti-air pollution activities paved the way for the increased efforts toward energy efficiency. Japan is often called a very energy efficient economy. In my view, this is because of air pollution control activities. The point is that if you try to reduce noxious emissions, you have to be careful with combustion technology. Otherwise we can't reduce much. Therefore such activities did pave the way for increased energy efficiency.

Dr Levine

It is hard for me to understand how the pollution activities led to efficiency. With the exception of noxious emissions, where combustion is very important, the same measures that improve efficiency pollution control may not work as expected. In many other cases, I see them going the other way. The energy efficiency efforts reduce capital expenditure if done correctly, but there is no guarantee that it will be done correctly. If it is, then capital investment in developing countries will be freed up as a result. You can then improve the standard of living through economic development and capital investments in pollution control.

Chapter Three

Activities of Environmental Protection

III.1 Overview

In this chapter are three papers dealing with the environmental problems from different perspectives, and hence three different initiatives toward environmental protection.

The first paper by S. Sudara deals with those problems faced in Thai society which are not necessarily the outcomes of economic development in Thailand alone. Rather, he argues, those have been brought about by the foreign capitals operating in Thailand. Of no less importance is the nature of the global economic system, dominated by the developed economies, which is not sustainable even in their perspective. Hence, he claims, activities for environmental protection should include the transformation of that system through an extended collaboration across the divisions of South and North.

On the other hand, R. Sprenger examines the importance of economic measures toward environmental protection, especially tax reform that includes environmental tax. Of course, as his paper discusses more in detail, there are pros and cons to this. By looking into past experience in the European Union, he proposes to create some economic incentives so that economic development and environmental protection can both be achieved even in the European context.

The last paper by M. Kurihara discusses the possibility for local authority to implement protective measures at grassroots. He explains what a city government can and cannot do, while also touching upon the importance of citizen participation, not just in the process of the implementation of those policies but also in the very policy making process itself. Especially interesting is the suggestion of a contract between the local government and those industries operating within the territory for the reduction of pollutant emissions. The last analysis will be one of the major areas for further studies, to make the protection more feasible, realistic and practical.

III.2 Environmental Problems and Activities for Protection in the Third World: the Thai Experience

S. Sudara

Professor, Department of Marine Science, Chulalongkorn University, Bangkok, Thailand

Abstract

This paper outlines the exploitation of colonized and undeveloped countries by developed countries. It looks at the trading in natural resources which went on up to the 1960s, widening the economic gap between developed and undeveloped countries and leading to resource depletion in the Third World. It also discusses the subsequent export by developed countries of heavy polluting and energy demanding industries to Third World countries, who sacrificed natural resources for economic development.

The paper concludes by outlining the attitude of developed countries towards environmental problems in the Third World, and the Third World's growing awareness of environmental management.

Human beings have always naturally used the resources available in their particular area. However, some countries develop more successfully than, and come to exert a degree of domination over, their neighbors. These countries inevitably consume more natural resources than their relatively weaker neighbors. This trend became particularly marked during the latter stages of the colonial era. Developed European powers actively sought new colonies with natural resources which they could transfer back home. The stronger European countries were thus able to develop even more rapidly and to widen further the economic gap between themselves and the colonies.

This resource exploitation was the primary cause of environmental degradation, but it was not confined to the colonies. Countries which had not been colonized, such as Thailand, also traded heavily in the export to developed countries of natural resources: timber, minerals and agricultural products in particular. Large-scale exports of raw materials inevitably led to resource depletion, of which the wholesale destruction of tropical native forests is a good example.

In the aftermath of the First and Second World Wars, increased industrialization led to even greater exploitation of these Third World natural resources, and to signs of environmental pollution in the developed world itself. By 1960 the industrialized countries were facing an environmental crisis from the effect of pollution both on human beings and on the natural environment. The industrial sources of some of this pollution, such as waste water discharges and smoke, were easily identified, and led to public pressure for remedial action. There was also invisible pollution, barely recognized, but potentially capable of intolerable effects. Rachel Carson's best seller *The Silent Spring* opened the eyes of the people at the grass roots, especially in the United States. In Japan, suffering and fatalities from industrial pollution were experienced in many areas. It seemed as if every kind of toxic pollution could be found in Japan. There were many confrontations between developers and citizens. Requests by local people for help were met by a widespread refusal to take any responsibility.

Towards the end of the 1960s many developed countries of the North began to attend to these problems and to clean up their environments. The 1972 Stockholm Conference marked the recognition of the global nature of environmental problems. Many nations, particularly those of the North, began to take control of, and to manage, their environmental condition. Many laws and regulations were enacted, and new techniques and management systems introduced, all designed to bring environmental problems under control. The 1970s could be considered the decade of the environment.

However, conflicts of interest could soon be clearly seen. The world was confronted by another problem—the energy crisis; the days of cheap energy were gone. The concern to create a better environment was left behind. At the same time, the emphasis on economic development was regaining its momentum. Decisions were being driven more by economic gain than by any concern for the environment. Meanwhile, the cost of environmental cleaning-up was becoming ever more expensive.

Ten years after the Stockholm meeting, an international conference in Nairobi in 1982 confirmed that the world's environmental problems had not been solved. Although some progress had been made in developed countries, there were still many issues in those countries on which no progress had been made. In developing countries, on the other hand, the extent of the environmental damage had grown, with little effort having been made to attend to the issue. The Nairobi conference agreed, among other things, that environmental problems were an issue of global rather than national responsibility.

Developed nations have found an easy solution with the export of heavy polluting and energy demanding industries to Third World countries without adequate environmental legal protection. Such countries, still struggling towards economic development, have usually welcomed any new industry. Faced with rapid expansion and high energy demand, natural resources were sacrificed. Tropical native forest was exchanged for hydro-electric development; energy production was expanded without proper planning.

Under such processes, Third World countries were driven into more serious environmental problems. Their natural resources were still largely demanded by the developed world. For instance, timber and wood chips for the pulp and paper industry were still in high demand and tropical countries met that demand. Polluting industries brought an increase in health problems. The local wealthy grew richer from their investments in developed countries, and continued to exploit local natural resources, while the poor became poorer from the loss of their livelihood as their natural environment deteriorated and, with it, their limited economic benefit.

At the 1992 United Nations Conference on Environment and Development, held at Rio de Janiero, Brazil, it was again made clear that the world's environmental concerns are global, rather than simply national, concerns. Nevertheless, the solutions proposed at Rio gave the impression that, because of their high population density and the ongoing depletion of their natural resources, the Third World countries were the cause of their own problems. The alarming rate of loss of tropical rain forest was considered to be the major cause of global climate change. Tropical nations were therefore asked to abandon the exploitation of their forests. At the same time, developed nations showed little concern about the need to reduce their own extremely high rate of resource consumption. The proposal to reduce the production of greenhouse gases failed to receive any firm commitment from most of those developed countries which contribute to the problem.

The environmental quality of Third World countries is still deteriorating. There is increasing evidence of the damage to both natural and artificial environments. Non-governmental organisations (NGOs) are fighting hard for environmental protection. More grass roots organisations are being formed to protect the land on which they depend. But conflict between economic development and protection of the natural environment is increasing. Multinational companies supported by investors from Third World countries continue to establish highly polluting and heavy energy consuming industries in developing countries to avoid the difficulties and expense involved in developed countries with their highly regulated standards of protection. Such corporations apply political pressure to establish and operate such industries in countries which have inadequate environmental protection. But the growth of NGOs and local groups will bring about greater confrontation both with developers and with authorities.

Governments in developing countries are making greater efforts to involve environmental studies in their decision making processes. New rules and regulations on environmental aspects are now being employed. Various government agencies have strengthened their management systems.

Despite this progress there are still many projects which are in strong conflict with environmental and conservation values. Economic development still often seems to dominate the minds of the decision makers. The growing NGO concern about these matters has led to the establishment of linkages between NGOs of developing and developed countries. Knowledge and awareness of environmental issues are being transferred to Third World countries. There is increasing pressure on developers to exercise care with their projects, and to adopt the same standards as are required in many developed countries.

Environmental awareness is growing because the influence of pollution on the environment can be recognized. There is now much evidence of the physical and mental suffering of local people caused by environmental degradation. But practices where the rich take from the poor and the strong take from the weak still exist. This is why more people's organisations are developing, to assist people to take care of their own future.

The environmental problems of the Third World are not yet solved, because contrary social and economic pressure are still often effective. But more careful consideration is beginning to be given before decisions are made. While there is still a long way to go, new techniques such as environmental impact assessments and risk assessments are now being used. Loan agencies, such as the World Bank and the Asian Development Bank, are now adding environmental impact assessments to their decision making processes. Even though not all of these assessments are accepted, and even though these practices are only beginning, at least it is a good start. Let us hope that the situation will continue to improve.

III.3 Initiatives for Environmental Protection in Germany

R. Sprenger

Professor, IFO Institute for Economic Research, Munich, Germany

Abstract

Germany has one of the most comprehensive environmental management policies and has acted on environmental issues earlier than most other countries.

This paper aims to describe the major environmental management issues in Germany and to summarize information about the economic and other instruments available for environmental management. These are: 'command and control', risk management and environmental impact assessment, information, education, voluntary agreements between affected parties or self-commitments of target groups, deposit/refund schemes, user charges, polluter charges, emissions trading and tradeable permits, liability for environmental damage, and subsidy schemes.

1 Introduction

Despite the fact that Germany has one of the most comprehensive environmental management policies and has acted on environmental issues earlier than most other countries, it is confronted with an immense task to reduce pollution and the flow of waste from industry and private households.

Since there are plenty of problems ahead for the policy makers and the target industries, there is an urgent need for regulatory reforms and effective policy instruments in the field of environmental management.

Economists and other analysts of government policy have long contended that the imposition of detailed rules may not be the best way to accomplish derived goals in environmental management. While such rules appear to go directly to the heart of things and guarantee the desired outcomes, they have been criticized for being economically inefficient, subject to excessive administrative and litigative delays, and demanding far too much detailed technical knowledge for the capabilities of federal and state agencies. For a variety of reasons—perhaps the most important of which is the growing concern with the increasing costs of environmental protection—this concern of academics has come to be shared by the federal and state governments.

Apart from tightening the traditional 'command and control approach' federal, state and local governments have developed and encouraged the application of economic incentives in environmental management. The specific (new) economic incentives include:

- user charges;
- voluntary self-commitments by industries;
- deposit refund system;
- subsidy schemes;
- environmental labelling;
- extended liability;
- tradeable permits;
- pollution charges.

All of these approaches are designed to introduce some flexibility into the means by which firms and households comply with environmental regulations, to produce significant savings in pollution abatement costs, and to remove existing barriers to technological innovation. They are also intended to make pollution avoidance and minimisation profitable.

Given this background, the objectives of this paper are:

- to describe the major policy issues in the field of environmental management in Germany;
- to summarize available information on the (planned) application of (economic) instruments for environmental management; and
- to analyse policy trends and to draw conclusions.

2 Major Environmental Policy Issues in Germany

The Federal Republic of Germany (FRG) has been one of the most environment-conscious of the European countries with the tightest environmental legislation and firmest enforcement. This is largely because:

- it has a high polluting, industry intensive economy;
- it has largely coal-based power generation;
- it was an early victim of the effects of acid rain on the national forests, pollution of the Rhine and the Chernobyl accident;
- it has a strong economy and secure, prosperous society allowing public attention to focus on quality of life issues; and (not least)
- of the urgent need to accelerate the cleaning-up and ecological modernization of the eastern states (former GDR) in the process of German reunification.

In order to reduce nitrogen oxides, public power stations must invest another DM 4 billion by the beginning of the 1990s.

Part three of the new Technical Guidelines for the control of air pollution (TA Luft) contains a concept for upgrading smaller power stations and virtually all industrial plants subject to the Federal Immission Control Law. The necessary retro-fits are estimated to cost a least DM 5 billion in investment outlays over the next four years. Moreover, the new ordinance on small combustion plants is targeting air pollution from private households, traders and farmers, who will have to invest annually DM 1 billion on the average for the modernization of their heating facilities.

In the area of waste water treatment, the amendments to the Water Management Law, as well as the waste water Charges Law, will bring new requirements for direct and indirect dischargers and thus release marked demand impulses:

With the fifth amendment to the Water Management Law, the requirements to be met when discharging waste water containing dangerous components have been considerably tightened.

The new regulation on the origin of waste water, which targets about 55 industrial and commercial sectors potentially discharging waste water containing dangerous components, defines technical requirements to be met when discharging waste water into rivers and other waters. The limit values for waste water components laid down in these provisions will in future also be the baseline for the approval of discharges into public sewers and sewage plants.

Against the background of increasing pollution of the North and Baltic Seas, the FRG adopted a 10 point programme for water pollution control in these waters. This programme requires a drastic reduction of phosphate and nitrogen discharges into waters and thus constitutes an enormous challenge to equip and/or retro-fit waste water treatment plants with non-pollutants. Another focal issue of this 10 point programme is the restriction of dangerous substances in waste water.

The amendments to the Waste Water Charge Law will also stimulate improvements in the field of waste water treatment. In operation since 1981, the waste water charges are levied on private and public dischargers according to the pollution content of direct discharges. The charges are lowered for those dischargers who meet the technical requirements, thus providing incentives for plants to surpass the required standards and the reduce their effluent discharges. The revenue from the charges is partly used for subsidizing additional abatement measures. The new legislation will result in a rise of the charges from the present DM 40 per damage unit to DM 90 per damage unit in the late 1990s. Nitrates and phosphates were also included in the elements subject to the new charge scheme.

Furthermore, for sewage drainage systems an average of approximately 1,200 km of new drainage pipes must be replaced. Financing incentives have also been created for this within the Federal Urban Improvement Programme.

Finally, the new Drinking Water Regulation stipulated by the corresponding EC directive will require denitration processes and plants in the field of water purification as a result of the stricter requirements concerning the nitrate content of water.

Waste minimisation and the control of hazardous waste have become the key themes of environmental policy in the FRG. Meanwhile Germany has one of the most comprehensive waste management policies in Europe. The term waste management in Germany includes the recovery of production of material or energy from waste (reuse and recycling of waste), the depositing of waste, as well as the necessary collection, transportation, storage and treatment. With the amendments to the Waste Disposal Act the FRG intends to transform traditional waste disposal into environmentally compatible waste management. As a result, noticeable effects, mainly with regard to waste disposal and waste recycling, will be seen. The private waste disposal contractors alone reckon with an investment volume of DM 4 billion to cover comprehensive recycling.

In addition to the Waste Avoidance and Waste Management Act in 1990, Technical Instructions for Waste Management (TA Abfall), which include general principles of waste minimization and technical requirements for the proper handling of waste, were released. Also, Technical Instructions on Special Waste (TA Sonderabfall) appended by three ordinances will stress waste avoidance, minimisation and recycling, and will give strict instructions for the disposal, chemico-physical treatment and incineration of all hazardous waste. The new legislative package will require waste generators to prove that waste cannot be recycled before a disposal permit is issued. Industry will be required to give detailed descriptions of their production processes etc. The aim is to identify and catalogue wastes to facilitate their recycling initiative.

The planned introduction of a progressive waste charge scheme will strengthen the incentives for waste avoidance and recycling in the FRG.

Apart from the reuse and recycling of newly generated waste, considerable expenditure will be required for the clean-up of so-called old deposits and contaminated factory grounds. According to estimates by the Federal Environmental Agency, the cost of coping with the pollution problems of the past will amount to a total of a least DM 17 billion over the next 10 years.

Mention must finally be made of noise abatement, for which the Federal Government alone will increase its expenditure (for road traffic noise abatement on federal highways) to DM 250 million annually.

The total investment required for the areas of environmental protection mentioned above has been estimated at between DM 290 and 375 billion for the western states of the FRG alone, over the next 10 years.

The present high level of pollution control spending in the western states of the FRG is not only likely to be sustained but is even to be increased quite significantly owing to the enormous task of cleaning up and modernizing the new eastern states of the FRG:

- In the area of air pollution, with its 5.2 million tons of $S0_2$ and 2.2 million tons of dust emissions per year, the former GDR ranked first in Europe regarding these pollutants. In some areas, air pollution has created immediate hazards to public health and damaged about 55% of the forests.
- Excessive use of pesticides and fertilisers in agriculture as well as the discharge of waste water without any treatment have been sources of serious

contamination of soils and water resources: the water quality of almost half of the waterways makes them unsuitable for the supply of drinking water.

- Vast scars to landscape and risks to groundwater bodies have been left by opencast coal and uranium mining.
- Almost half of the household waste was dumped on 'wild' landfills.
- The total number of contaminated sites has been estimated at more than 30,000.

The FRG government has meanwhile launched several investment and financing programmes to gradually improve the environmental quality in the new eastern states to reach a 'German environmental union' by the end of the 1990s. According to various estimates, this will require environmental investment in the order of DM 200 billion at a minimum.

3 The Role of Instruments in German Environmental Policy

Apart from tightening regulations, federal, state and local governments have developed and encouraged the broader application of market-oriented approaches to environmental management. The specific regulatory reforms include:

- risk management and environmental impact assessment;
- information;
- education;
- voluntary self-commitments by industries;
- deposit refund systems;
- user charges;
- pollution charges;
- emissions trading;
- extended liability;
- subsidy schemes.

3.1 Command and control

Environmental policy in Germany, as in all industrialized countries, is underpinned by legislation which defines the basic regulatory framework within which policy instruments are applied.

New basic environmental legislation has been passed recently reflecting a perceived need to strengthen existing legislative provision, partly in response to EC directives, partly reflecting the growing awareness of the need to strengthen environmental control.

Legislation with respect to pollution in Germany since 1970 is perhaps best described by the phrase 'command and control'. A pollution control agency—national, state or local—gives the 'command': a numerical limit on the emissions of a particular pollutant from a particular point, such as a pipe or vent from an industrial process. The owner or operator of the pollution activity must then 'control' emissions from that point to comply with the command. Under the command and control approach, German pollution control agencies have sought better environmental quality through rules which apply specific, uniform emission limits—generally based on a known, feasible control technology—to every emission point within a regulated process.

The command and control approach produced large gains in many areas, because it dealt with sources that were easy targets and where initial costs of control were relatively low. Despite added pollution from industrial growth, air and water quality in Germany have improved significantly since the early 1970s.

But the traditional approach, despite its successes, also has serious defects. First, rules are general statements that can't fit every case. They leave plant managers little room to adjust broad requirements to particular situations. And they treat similar processes equivalently, ignoring both different impacts on environmental quality and public health and the chance that equivalent emission reductions might be more easily obtained from other processes in the same facility or area. Because some sources are under-controlled by this approach, while many are controlled more than they need be, the result is excess costs—a powerful reason for industry to try to limit its obligations by resistance or delay.

Perhaps most importantly, the present system discourages innovative control approaches. A source that installs known controls can generally count on avoiding enforcement and penalties regardless of these controls' cost-effectiveness. A source that develops more cost-effective or efficient measures receives no reward or credit for doing so. Worse still, an innovative firm risks making itself the target for extra regulation, since it may have shown that it or its industry can do more. For most firms, it is simply not 'profitable' to invest in innovative efforts to do more than the minimum; only increased innovation can guarantee improved ambient quality at reduced—rather than increasing—cost. For example, to meet national ambient air quality standards in areas with continuing pollution problems, it would be necessary to squeeze more emission reductions from sources already regulated—at steeply increasing costs per unit of reduction. It would also be necessary to seek emission reduction from small, previously unregulated sources—resulting in a more complex and expensive regulatory process which returned fewer unit reductions for each action taken.

3.2 Risk management and environmental impact assessment

These are two instruments which are consistent with the tendency of placing responsibility on industry's shoulders.

The need to address the management of risk is gaining support at both national and EC levels. Both individual and societal risk are considered, but the risks to society and the environment posed by industry are beginning to take precedence.

In the FRG integrated safety analysis is required; legislation on limits in volumes, distances and time of contamination are expected in 1990-1995; and the stricter application of safety analyses in existing companies has been passed (new investments, constructions and plants are already affected).

3.3 Information

The governmental instruments are only effective to a limited extent if they do not meet with the understanding for polluters as to the necessity of the environmental protection measures. If environmental policy is to be successful on a permanent basis it is essential that those involved accept the goals of environmental policy and are prepared to actively contribute to their achievement. Environmental awareness is consequently a significant prerequisite for the success of an environmental policy.

The federal government considers necessary the further development of environmental awareness by providing rapid, comprehensive and precise information on the environmental situation and on possibilities of environmentally sound behaviour.

There are several ways in which information can be sought, acquired and used, including, for example, the public right to know, environmental labelling and environmental auditing.

Public right to know

While many environmental and energy campaigns have been based on exhortation and advertisement, this has rarely been systematic, and current thinking seeks approaches which can be harnessed to change behaviour or consumption practices. As yet this approach is in its infancy in Germany.

There is an EC directive (the Seveso Directive) requiring industry to provide information on a number of potential hazards, which is currently being implemented at national level.

Another new EC directive requires freedom of access to information on the environment held by public authorities. The information is to include all data, factual and legal, concerning:

- the state of, measures to preserve, protect and improve the quality of and any changes therein, water, air, soil, fauna, flora and natural sites, plus measures designed to prevent damage and repair any damage caused;
- public and private projects likely to damage the environment or endanger human health and plant or animal species, in particular as regards emission, discharge or release of substances, living organisms or energy into water, air or soil, and the manufacture of dangerous products or substances.

Under this directive, trade and industry secrets will be protected by restricting the right of access to information where it would affect, *inter alia*, such secrets. At a national level, the role of pressure groups is proving especially important in obtaining freedom of access to environmental information.

Environmental labelling

'Eco-labelling' can also be considered an economic incentive or at least a facilitating mechanism for the reuse and recycling of waste. Consumers wish to be assured of the environmental credentials of the products being purchased. Therefore, green consumerism to promote recycling and waste minimisation can only work if consumers are able to judge what is and is not either recyclable or based on secondary raw materials. Given the complexity of environmental issues and the unproven scientific literacy of the general public, it is clear that 'eco-labelling' schemes can greatly improve their ability to make that judgement.

Against this background 'eco-labelling' has become an important aspect of Germany's environmental policy. Over 4,000 products in 70 categories now carry the 'Blue Angel' label, and it has been observed as achieving one of its main objectives. This is to provide an incentive both to industry, to rethink products and to develop recyclable and low waste products, and to consumers, to express their preference by choosing recyclable products. There is also evidence that public procurement has assumed a pioneer role in recycling or the purchase of recyclable products with an 'eco-label'.

However, the future role of eco-labelling to discriminate in favour of such environmentally compatible products in government purchasing and private consumption will largely depend on an EC initiative to develop an EC-wide eco-labelling scheme.

The opposite of eco-labelling appears to be the mandatory labelling of certain products, eg hazardous substances, to make sure that they are returned, treated and disposed of separately. This type of labelling, which is requested by the new Waste Avoidance and Waste Management Act, will probably create a disincentive for the marketing of these products or, in other words, create an incentive to withdraw such products from the market.

Environmental auditing

A recent development in the EC is for companies to assess the impacts of their activities on the environment and take mitigation steps. Many companies, particularly suppliers of consumer products and retailers, have carried out 'green audits' to identify environmental vulnerabilities and assets. The EC proposal stimulates a voluntary commitment by industry to conduct an environmental audit of their activities.

3.4 Education

Information and education clearly complement each other. Education is necessary to enable industry, consumers and government bodies to interpret and act upon information made available. Environmental labelling is only helpful if consumers know what the implications are. It is equally important that monitoring authorities are able to use information generated by environmental audit.

An effective long-term education programme on the environment is necessary if ideas and actions are to be given a new direction and the environmental awareness of the population is to be consolidated in the future. Both federal and Länder governments have particular responsibility here within their field of competence. Education about the environment is not only aimed at conveying skills and knowledge for practical industrial activities, but rather at influencing the attitude of the individual to intensify their awareness of responsibility to the environment.

The federal government thus attaches great importance to education about the environment. In the future it will work towards intensifying the integration of activities concerning the environment in all areas of education.

Environmental education in Germany is offered in a number of ways by different groups:

- Governments have used education and exhortation to change behaviour patterns for many years, through such campaigns as the energy saving campaigns of the early 1970s following the first oil crisis, designed to change consumer behaviour. A series of free publications has been recently published on energy efficiency by the business sector, such as retail and other services.
- Governments also initiate campaigns targeted at certain policy issues, for example the Waste Management Programme uses a combination of technical assistance, demonstration projects and training to encourage industry to reduce waste at all stages of production. There are several business guides made available to industry which offer practical advice and case study experience on waste minimisation and recycling.
- Lobby groups have launched educational campaigns directed at consumers and retailers. For example, environmental groups in Germany initiated a campaign to reduce the use of tropical hardwoods for furniture, using a mixture of educating retailers and distributors, campaigns to promote alternatives, and campaigns aimed at government bodies to encourage funds to promote forestry management.
- 'Green' newspapers and supplements to the press, and 'green' politics play a role in providing understanding and educating people to be able to make informed choices.

As environmental management moves away from single plant process control to product management and environmental quality, education and information are likely to become more important. This pattern also reflects the tendency to make industry and consumers increasingly responsible for their own environmental management through audits and information requirements at industry level, or household waste sorting in the domestic sector.

3.5 Voluntary agreements between affected parties or self-commitments of target groups

In most cases, legislation and enforcement under the traditional command and control scheme is a time-consuming and burdensome process. Therefore, both regulators and target industries have a vested interest in means that could provide quicker, unbureaucratic, more flexible and cost-effective solutions. In this context, voluntary self-commitments and pledges by target industries or agreements between governments and industries have often taken the place of regulation (see Table 1).

This approach has also been adopted in several areas of waste management, eg such as reuse and recycling quotas for glass beverage containers, paper, cans and other scrap metals, and the acceptance and guaranteed disposal of returned products, such as products involving CFCs or mercury batteries. Other agreements refer to the clean-up of landfills contaminated by industrial wastes. The most recent 'voluntary' self-commitment refers to the 'dual waste disposal concept' that aims at reducing household packaging waste by 25 to 30% in weight or by 45 to 50% in volume and recycling as much of it as possible (see Table 1).

However, these voluntary agreements haven't always been a success. In particular, the agreement about the reuse of beverage containers failed, thus requiring more effective government intervention. Similarly, the cleaning-up agreements will not solve the immense problems in this area. Therefore, for voluntary agreements, self-commitments and pledges to be successful the targets must be clearly defined, the number of players involved must not be too large and the threat of tough government intervention in case of failure must be real.

Voluntary agreements can be useful when involving large companies or well organized sectors of industry, while free behaviour of companies in fragmented sectors can be directed with much greater difficulty. Even the sectoral associations of entrepreneurs cannot be fully responsible for their associated companies' choices. If the industrial structure is very fragmented as it is in many German industries, other policy instruments are more reliable than voluntary agreements.

Even in the case of large companies some problems exist. The public administration's ability to control and enforce execution of investment programs is crucial to effectively achieving the agreed reduction in pollution. Even if sanctions cannot be used in the case of a voluntary agreement, the policy maker should have ways to enforce the execution of the programme if necessary.

If this is not possible, then any unforeseen external event may reduce the management's determination towards introducing of environment-oriented investments, substituting of raw materials, changing products or production processes. Pessimistic economic forecasts or negative news about the demand of some products, as well as changes in property or management, may induce change or straight cancellation of previous plans.

3.6 Deposit/refund systems

Germany has a long experience with voluntary deposit/refund schemes in the area of glass bottles. However, this self-commitment of the food and drink industry hasn't proved to be a success. In the period from 1970 to 1987 the reuse of glass bottles has seen a significant decrease, going down from 90 to less than 75%. This negative trend was expected to be reinforced and accelerated by the planned introduction of non-returnable PET bottles by Coca-Cola. To prevent a further increase in the use of non-returnable bottles the German federal government has enacted legislation on plastic beverage containers that involves a mandatory deposit of DPF 50 on such containers and makes traders accept the return of plastic bottles.

To further reduce the flow of waste, Germany has recently adopted obligatory deposit/refund schemes for all types of beverage containers, and for containers and packages for detergents, cleansing agents and emulsion paints, which oblige shopkeepers to accept the return of such containers from 1993.

3.7 User charges

User charges, ie charges for the collection, transport, treatment and/or disposal of waste and waste water, are long-standing instruments in environmental manage-ment, though they have not been seen as economic instruments until recently.

Traditionally, user charges are the means of achieving full cost recovery for waste and waste water disposal or treatment services by local communities. User charges for waste disposal, for example, usually would include a fixed service fee, a volume based charge derived from the volume of waste that is collected, and a supplementary charge for wastes that need a special treatment. The charge schemes are administered by local authorities, and the level of costs met by the charge varies from place to place.

There are a number of problems that can occur with user charges for waste disposal: there is a wide variation in the level of coverage and the basis for charging; it is sometimes difficult to balance the level of charge and the expenditures which can cause local problems; the charging structures tend to be closely linked with municipal financing structures; and where the charges don't cover full costs of the services they can hardly stimulate waste minimisation.

However, the use of full cost recovery charging as a waste management instrument is becoming more and more important. Local authorities are promoting full cost recovery schemes, thus making waste disposal more expensive and encouraging waste minimization in households as well as low waste technologies in the industry.

3.8 Pollution charges

Pollution charges, levied on the pollution emitted but not related to the costs of treatment or disposal, have been in use in Germany as an economic instrument to encourage better waste water treatment. In operation since 1981, the Waste Water Charge Law has been amended three times and the parameters of the charge, the charge tariffs and the conditions for exemptions have been updated.

Against the background of the experience with waste water charges, the Federal Government recently decided on a waste charge. The new waste charges are to be levied on waste generators and on the private or public operators of waste treatment plants according to the volume of waste generated. In case there is a need for disposal of residues after waste treatment a supplement for waste disposal must be paid. The new legislation will classify the wastes requiring the highest charge for the generation and disposal of hazardous waste (DM 100 per ton or parts of 1 ton) and applying the lowest tariff to household refuse (DM 20 per ton). The charges will be raised by 20% every two years through to the year 2002. Waste generators will be able to offset waste treatment or disposal investment against charges while the equipment is being installed if the investment is to reduce waste generation by at least 50%.

The revenue from the charges is intended to cover the administrative costs of the system as well as providing support for research and development, pilot projects, promotion of low waste and other waste avoiding technologies. Forty percent of the expected annual revenue (DM 5 billion) will be used for the clean-up in eastern Germany.

Even though it is premature to make any prediction about the probable quantitative effects of the new charge scheme, it will without doubt have a supplementary incentive effect to minimize or even avoid waste generation.

Plans to introduce a national pollution charge on $C0_2$ emissions have been cancelled due to the EC initiative for an EC-wide $C0_2$ energy tax.

3.9 Emissions trading and tradeable permits

Emissions trading rests on the concept of determining a total permittable level of pollution for a given area and allowing polluters to buy, sell or trade permits within that overall limit. This approach is seen as allowing economic efficiency gains as a result of the flexibility given to the target firms to comply with given emission standards.

Germany has adopted parts of the United States emissions trading concept in the field of air pollution control. The so-called bubble and offset approaches were introduced into the Federal Emmission Control Act in 1985. These options were already referred to in the Technical Instructions on Air Pollution Control in 1986 by providing the possibility that, in specific areas, existing installations may deviate from regulatory standards if more advanced measures were taken by other installations located in the same area, thus achieving a higher level of overall air quality for this specific area. However, these options have remained almost insignificant to date because of the highly restrictive conditions placed on their application. In the recent amendment to the Federal Emmission Control Act the legal requirements to be met by bubble and offset proposals were, on the basis of the experience gathered so far, significantly extended in line with the pollution control objectives of this law.

Apart from air pollution control, Germany has had some experience in emissions trading in the area of waste management. The state of North Rhine-Westphalia passed a new waste disposal law in 1989 including provisions for tradeable permits in the field of waste disposal. According to the new law, a waste management company may only take part in the business of disposal of hazardous wastes if it holds a licence or permit for this type of activity. These permits can be acquired from the State Waste Water and Waste Agency, which controls the issue of permits in accordance with the waste management objectives of the state. The revenue from issuing the permits will be allocated to a fund (Abfallentsorgungs—und Altlastensanierungsverband NRW) which takes special responsibility in the waste disposal and the cleaning up of contaminated sites. This fund will also operate its own waste treatment and disposal plants in order to ensure that bottle-necks in waste disposal do not arise. Other activities include research, planning and the design of adequate means to promote waste avoidance and minimization.

It has been argued that this approach is still far from real emissions trading, since the issuing of permits only requires fee payments and is not based on a market-oriented auction. Another problem arises with the EC internal market rules that would allow waste generators to export their wastes, leaving the licence holders with an uncertain economic future with respect to the revenue from domestic waste treatment.

3.10 Liability for environmental damage

Liability has always been some sort of catch-all instrument for environmental protection in the FRG where and when responsible polluters could be found and required to recover the necessary clean-up costs. The growing interest in the application of strict liability in the FRG reflects both the sense that the polluter must be held responsible for their environmental damage and the United States experience where the Superfund Acts have proved a useful means of compensation and an effective incentive for waste minimization. Strict liability under the new German law implies that liability for environmental damage is retro-active and very strict, imposing joint and several liability for soil and water contamination by specific plants generating hazardous wastes or for damages from accidents as well as from authorized operation of such plants or waste treatment plants. The new legislation also includes a mandatory insurance against liability risks thus creating an economic incentive in the form of varying insurance tariffs according to the individual risk potential of plants and operations.

Strict liability in the area of waste management will be stengthened by the EC Directive on Strict Liability for Transfrontier Shipment of Wastes. Another proposed EC directive on civil liability for damage caused by waste would:

- impose strict liability on the producer of waste until waste was transferred to an authorized disposal site or facility;
- establish liability for damage including death, personal injury, damage to property, and damage to the environment;
- enable legal action to be instituted to require restoration of the environment or to recover costs.

3.11 Subsidy schemes

When it comes to cleaning up the environment and to stimulating recycling and waste minimisation, the Federal and State Governments are turning to the 'stick and carrot' approach. Command and control, user and emission charges, or strict liability are mostly of the 'stick' variety. But subsidies—the 'carrot' which governments dangle before industry—have a large role to play too. Governments justify their use by arguing that they speed up and provide incentives for industry to invest in waste disposal and to develop low waste technologies or recyclable materials and products, and that they compensate industry partly for the costs of meeting environmental regulations earlier than the competition abroad.

By raising revenues through charges, government can use both the stick and the carrot at once, as has been the case with the former charge on lubricating oil to subsidize the collection and reprocessing of the relevant used oil, and the waste water charge to subsidize the upgrading of waste water treatment facilities.

There are many cases in the FRG where subsidies are used in the area of environmental protection. The main subsidy schemes in use in the FRG include:

- tax allowances (eg for the purchase of private cars with catalytic converters and the consumption of unleaded petrol);
- grants or soft loans for private investment (by two government owned banks: KfW and DAB);
- assistance for research and development of cleaner technologies or for pilot and demonstration projects, and
- financial assistance for local authorities through grants and soft loans.

In general, the use of subsidies to assist private investment in pollution control or to cross-subsidize municipal waste and waste water disposal where the full costs have not been passed on to the user ('hidden subsidies') is contrary to the polluter pays principle. Another argument against subsidies is that, where the revenue is drawn from general taxation, this is inequitable as some taxpayers do not benefit. Since the majority of existing subsidy schemes are explicitly linked to regulations, they hardly induce activities going beyond those requirements.

There is an increasing subtlety in the use of subsidies, eg the German Waste Water Charge Law offers reductions in charges for firms installing state of the art technology. It also offers drawbacks on payments of charges while the equipment is being installed. Thus the industry is effectively subsidized by not having to pay the pollution charge associated with its discharge even before pollution control equipment has been installed.

Moreover, most of the schemes focus on 'end of pipe' investment and therefore do not give an incentive for pollution avoidance or minimisation.

To sum up, there is an urgent need to focus subsidies more on clean, low waste and recycling technologies and on research and development with respect to recyclable or more environmentally compatible products.

4 Other Pressures on the Private Sector

In addition to these public policies and instruments, pressure on the polluters and waste generators to improve their environmental performance comes from various groups of the society, as Figures 2 and 3 show:

- Industrial and commercial associations that often develop voluntary agreements, (eg on eco-labelling, deposit refund systems, recycling targets) to avoid tight, technology-forcing and less cost-effective regulations;

- Customers who, for commercial or ethical reasons, wish to be assured of the environmental credentials of the products being purchased. For consumer products, this is the phenomenon of 'green consumerism';
- Members of the public acting on the basis of information mostly supplied through the media. They are motivated by general altruistic concerns, often expressed via pressure groups like BUND (Bund für Umwelt—und Naturschutz in Deutschland), Greenpeace and other environmentalist movements, or by self-interest, usually in the case of local residents in the areas where it is proposed to locate new incineration plants or landfills—the so-called NIMBY (not in my backyard) syndrome;
- The financial sector, including insurance companies, providers of capital, green and ethical unit trust groups, and potential/actual merger and acquisition partners, all of whom need to know the target company's current or potential environmental liabilities;
- Employees, actual or potential, as well as their shop stewards, who want to be assured of the environmental credentials of the company for which they work;
- Trade unions, who want to make sure that the industry for which their members work won't have to lay off their members as a consequence of non-compliance with the environmental legislation.

5 Trends and Conclusions

The changing perspective in environmental policy approaches appears to be reflected in a broad shift in policy emphasis away from traditional command and control instruments in favour of economic instruments and market mechanisms, information, liability and education.

The underlying shift is more complex. The shifts in perspective in environmental policy have meant that, to a growing extent,

- government authorities perceive that direct emission controls on industry have reached the limit of effectiveness. Regulations for specific emissions such as $S0_2$ and $N0_x$ and the installation of pollution control equipment have meant that even if targets have not yet been reached they soon will be; the mechanisms are now in place. These direct control mechanisms have been highly effective, but it is felt that the approach has gone as far as it can;
- there is an increasing concern with individual products or waste streams—such as plastics and PVC, dioxins, and heavy metals which are used to some extent throughout industrial processes and whose control is more complex than emissions which can be resolved by a straightforward investment in cleaning equipment. The control of such products affects the whole production cycle, the use of the materials and the search for substitutes;
- the number of chemicals or materials for which environmental controls are sought is steadily expanding; regulation by product becomes an increasingly unwieldy task.

The combination of these two patterns—greater numbers of damaging substances being identified and recognition of the need to switch from end of pipe clean up to products and processes—has stimulated the search for new approaches, and the emphasis has shifted from imposed standards to a requirement on industry (and other polluters such as waste producers) to examine their own practices and adopt cleaner, better practices.

This pattern has been strongly established in the new regulatory instruments which are in use, such as risk management and environmental impact assessments. The information and education approaches force industry to publicise the way in which it impacts on the environment. The ultimate instrument is to force polluters to accept liability, retro-actively, severally and corporately.

These new regulatory patterns are reinforced by the use of economic instruments which intervene in the market and by giving price signals through imposing charges on the pollution in discharges or taxes on environmentally damaging chemicals (such as carbon) to encourage polluters to change their behaviour and to internalise their externalities (see Figure 3).

The experience in the FRG with respect to the use of economic instruments can be summarized as follows:

- There is no such thing as a cure-all among the economic instruments for environmental management. In fact, depending on the pollution or waste problem, various specific market-oriented approaches have been implemented.
- Economic instruments are not operated in isolation; they are generally used within the regulatory framework.
- There are many cases where specific economic instruments are applied in conjunction with other instruments (mixed approaches). Combinations of an economic instrument and direct regulation are quite common, either to reinforce such regulation, or to provide for necessary funds.
- Economic instruments may precede adjustments of existing regulations or a new regulation as in the case of voluntary agreements. Major direct regulations take time to be designed or adjusted, particularly if international bodies are involved. Temporary measures with a financial character can speed up compliance with direct regulation in advance of their actual implementation.
- Sometimes a combination of economic instruments can be fruitful. Examples are eco-labelling and deposit/refund systems (for instance for packaging) and user charges on non-returnable components.
- Even though it is often difficult or even impossible to evaluate or to predict the positive effects of economic instruments, especially in the case of a mixed approach, there is a growing interest in the broader application of economic instruments for environmental management in the FRG.

Figure 1
Disposal of product packaging in the Federal Republic of Germany

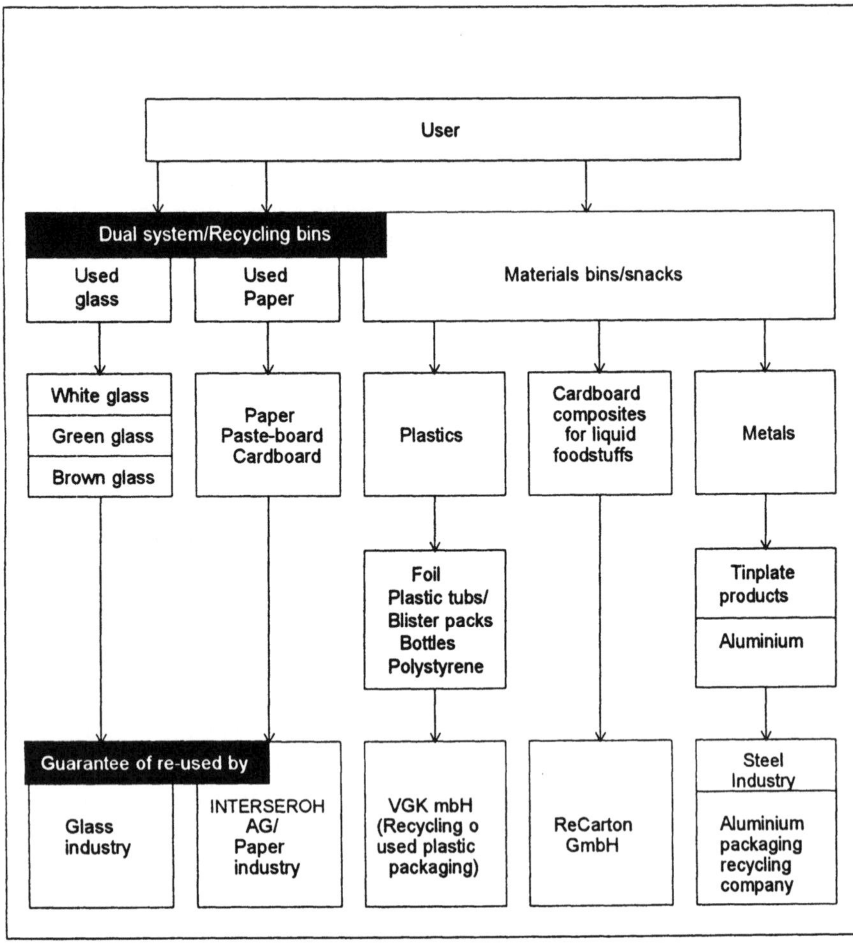

Figure 2
Pressure on polluters today

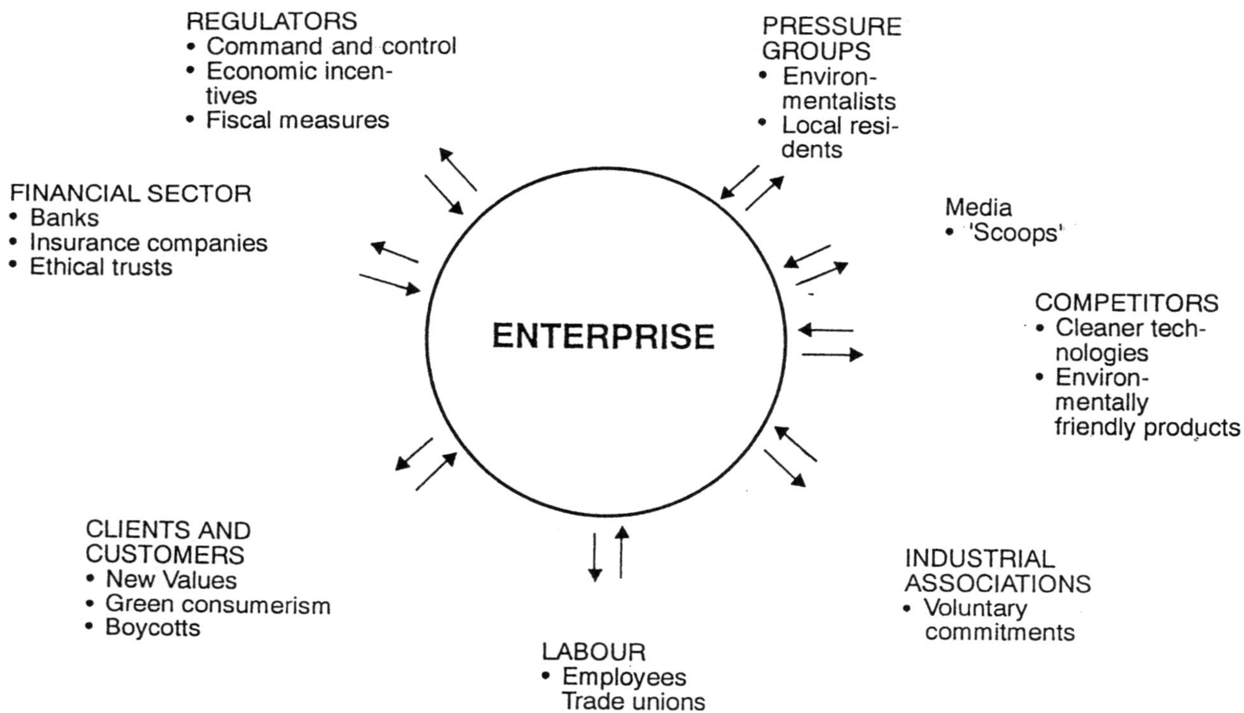

Figure 3
Vision of the evolution of driving forces in environmental protection

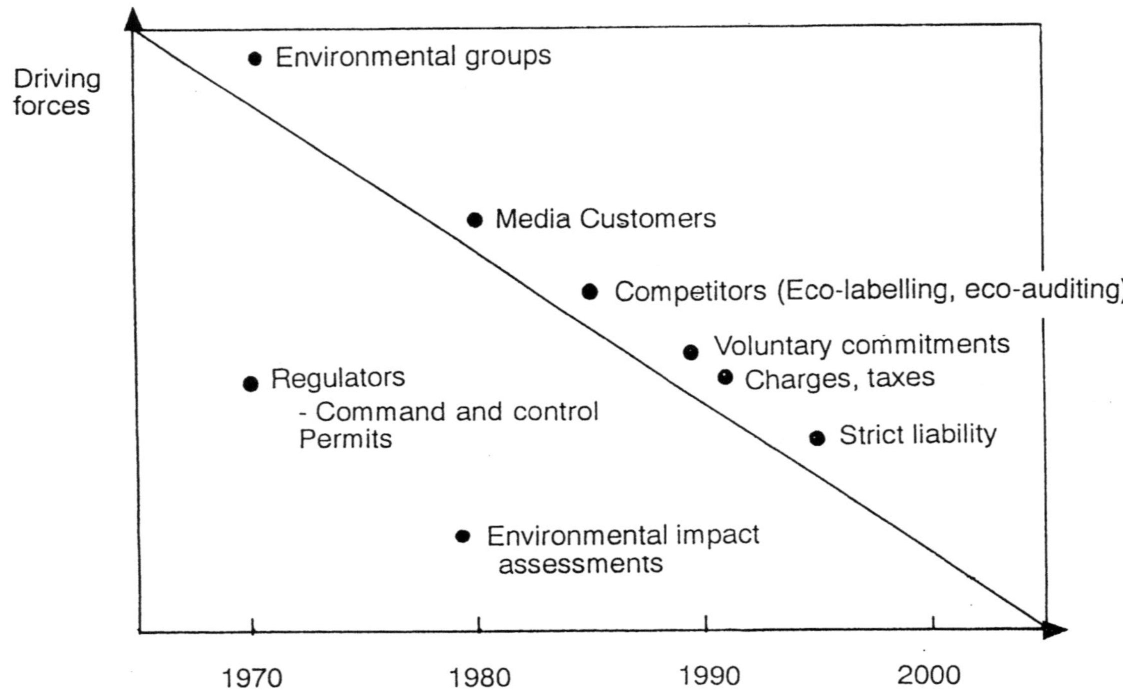

Table 1

Voluntary agreements and undertakings by industry for environmental protection

- Voluntary agreement on completely phasing out asbestos in building materials used above ground by 1990 (1984)
- Voluntary undertaking by the paint industry to reduce the content of solvents and heavy metal compounds in lacquers and paints (1984)
- Voluntary agreement by the paint industry to reduce environmentally harmful substances in underwater paints for boats (1986)
- Voluntary agreement to improve technical washing and cleansing installations (1986)
- Voluntary control of old substances in the chemical industry (since 1985 control of 60 substances of particular importance)
- Code of conduct for the chemical industry for the exportation of dangerous chemicals (1986)
- Reduction in the use of chloro-fluoro-carbons as propellant in aereosols (1975)
- Voluntary undertaking to discontinue the use of penta-chlorophenole in wood preservers
- Programme for the reduction of the ammonium content in effluents of the chemical industry (1986)
- Numerous voluntary agreements in the field of detergents and cleansing agents
  - Marking of detergents containing enzymes (1971
  - Dosage recommendations for detergents (1973)
  - Limitation of the phosphate content in detergents/washing powders
  - Submission of the basic composition of drugs to health authorities in order to help prevent infant poisonings (1975)
  - Recommendation on the limited use of nitrile triacetate (NTA) in detergents/washing powders (1984)
  - Agreement to affix a warning notice to packages of dishwasher detergents (1980)
  - Voluntary agreement on the use of child-proof packages (1985)
  - Voluntary agreement on household detergents containing hypochlorite (1985)
  - Voluntary agreement to end the use of certain softening agents (acylphenolethoxylates — APEO) in detergents/washing powders and cleansing agents (1986)
  - Voluntary undertaking to supply data on the environmental compatibility of detergents/washing powders and cleansing agents (1987)
  - Voluntary undertaking to discontinue the use of volatile chlorinated hydrocarbons in detergents/washing powders and cleansing agents (1987)
- Voluntary measures to reduce waste from drinks packaging (1977)
- Increased efforts to recycle waste paper, glass and tin-plate (1977, 1982)
- Vermittlungsstelle der Wirtschaft für Altlastensanierungsberatung (1985) (VAB — a contact for firms seeking counselling on the redevelopment of hazardous sites)
- Advance compliance with statutory requirements to reduce emissions from large combustion plants (especially in North Rhine-Westphalia and Baden-Württemberg)
- Voluntary undertaking by manufacturers of gas-fired household appliances and equipment to improve the energy efficiency of their products (1980)
- Voluntary undertakings by manufacturers of electrical household appliances to cut the power consumption of such appliances (1980)
- Voluntary undertakings by the automobile industry to cut the fuel consumption of passenger cars and commercial verhicles (1981)
- Information campaign by the mineral oil industry on unleaded petrol (1986)
- Voluntary measures offered by industry to reduce waste from drinks packaging (1987)
- Voluntary programme of the chemical industry to reduce the use of chloro-fluorocarbons in aerosols (1987)
- Voluntary programme of the lighting industry and the energy industry for the recycling of fluorescent tubes (1987)
- Safety programme of the chemical industry, concerning plant and equipment for the handling of water pollution substances (1987)
- Voluntary programme of manufacturers of electrical household appliances, concerning the disposal of refrigerators and freezers (1988)
- Voluntary agreement on the disposal of used batteries (1988)
- Voluntary agreement on the production and marketing of finger paints (1988)
- Voluntary undertaking by the chemical industry to phase out the production of all CFCs listed in the Montreal Protocol (1990)
- Voluntary agreement on the labelling of drinks packaging
- Voluntary agreement to take back batteries (button cells) and to reduce the mercury content in alkali-manganese batteries
- Agreement on the production and marketing of finger paints
- Restrictions on brominated diphenyl oxides in plastics
- Reduction of volatile chlorinated hydrocarbons in effluents of the chemical industry

III.4 Policy Coordination Between Citizens and Local Government for Environmental Protection

M. Kurihara

Environmental Policy Officer, Kawasaki, Japan

Abstract

The city of Kawasaki has had serious industrial pollution and other environmental problems, many brought about by sharp population increases and urban development. It has developed environmental regulations to deal with these in the past, but localized handling of the problem is no longer enough. The city is now looking at new approaches to deal with these problems in the twenty-first century. This paper outlines the administrative structure of the city's Basic Ordinance for the Environment.

Introduction

Kawasaki is a city with a population of 1.2 million, located between metropolitan Tokyo and the city of Yokohama. The city has aggressively pursued a policy of environmental regulation in order to cope with serious industrial pollution and a variety of other environmental problems. These problems have been aggravated by the city's geographic location, concentrated industrial activity and sharp population increases.

Given the latest environmental problems it faces, the city of Kawasaki aims to develop a new environmental policy with a new perspective as it approaches the twenty-first century. This article reports on the history of the city's environmental regulations and the direction and problems associated with Kawasaki's new policy.

1 Environmental Contamination Attendant on Industrial Activities

During the period of rapid industrialization in the 1960s, the contamination of the environment was first recognized formally as 'industrial pollution'. With the expansion of productive activities centering on the heavy machinery and chemical industries, air and water pollution worsened with a particularly serious impact on public health, giving rise to respiratory and other diseases. Accordingly, solving the problem of industrial pollution became an urgent and grave concern for Kawasaki.

At that time, the pollution control measures applicable in the city included the national Atmospheric Contamination Prevention Law and Noise Control Law as well as the Kanagawa Prefecture Pollution Protection Ordinance. These measures, however, were not fully effective in curtailing industrial pollution within the city, with its heavy concentration of industrial activities.

Accordingly, Kawasaki embarked on its own anti-pollution program in 1971, and in March 1972 it passed the present Pollution Control Ordinance of the City of Kawasaki. Paralleling the environmental quality standards stipulated by the Public Nuisance Countermeasures Basic Law, this ordinance established the city's own environmental quality targets and quantity controls on large area pollutants.

Another by-product of the high rate of industrialization was the rapid disappearance of valuable green tracts of land within the city limits. This was the result of vigorous residential development activities centering in the northwestern part of the city. A citizens' movement led to the passing of the Ordinance Relating to the Conservation and Restoration of Nature in the City of Kawasaki in October 1973, recognizing citizens' rights to apply for the designation of an area as a green zone.

Large-scale development activities throughout the city had also led to an undesirable imbalance between the city and its surrounding environment. The city adopted therefore, in October 1976, the first environmental impact assessment ordinance in the country—the Environmental Impact Studies Ordinance of the City of Kawasaki. This ordinance was designed to institutionalize the measures to be taken by developers prior to the initiation of development projects.

2 Current Environmental Problems in Kawasaki

The implementation of the city's environmental policy, which is centered on these three ordinances, produced a measure of localized success in controlling industrial pollution and regulating specific development projects. However, as the country's socio-economic activities undergo both qualitative and quantitative changes, there remain many unresolved environmental problems, even as new ones arise.

Typical of these are problems that involve the so-called pollution of the urban living environment. This includes air pollution by automobiles, groundwater contamination by residential waste water and waste disposal, as well as pollution by harmful chemicals used in high technology industries. The city is being forced to cope with the problems of creating a comprehensive urban environment designed for qualitative improvements in living conditions while providing adequate urban amenities and preserving a healthy global environment.

Solving these problems requires that people regain their awareness of the importance of the environment and better understand the fact that human activities have a significant impact upon it. The city government not only has to lead people with policy measures, but also provide support to citizens' efforts which aim for a better understanding of environmental issues.

3 Development and Promotion of a New Environmental Policy Regime

Inasmuch as more recent environmental problems have been closely intertwined with the city's socio-economic activities and the daily life of its citizens, localized handling by the vertically-oriented legal and administrative systems of the past is no longer adequate. For this reason, the city promulgated in December 1991 and put into force in July 1992, the Kawasaki City Basic Ordinance for the Environment, which takes into account expected future changes in the city's urban structure, its economic activities, and the life-styles of its citizens. This ordinance has the following five features.

- First, the ordinance defines the objectives and ideals of the environmental policy, taking into account its fundamental nature and its future prospects. It then identifies five principles which are indispensable for the realization of these ideals, including the comprehensiveness of policy measures and scientific predictability. These ideals and principles will help integrate the vertically-oriented regulations of the past and systematically induce the environmental policy as a whole to develop in a new direction.
- Second, the ordinance specifies the nature and content of the Basic Environment Plan and the system managing its progress. In the past, many local governments developed environment management plans. The great majority of these plans, however, serve merely as general guidelines for administration, and the regimes for implementing and managing their progress appear to be insufficiently developed, to say the least. Taking this potential problem into account, the latest ordinance has accorded a legal basis to the Basic Environment Plan and established the annual reporting system and Environmental Policy Advisory Council as an ordinance mandated process to assure faithful execution of the plan.
- The third feature of the ordinance involves the establishment of the Environmental Coordination Meeting as an internal administrative organ designed for the overall coordination of the environmental policy. Along with this, a system of environmental impact studies was adopted in order to assure thorough consideration of the environment in every environment-related policy measure and project of the city.
- Fourth, the ordinance supports the establishment of the Environmental Policy Advisory Council, whose function is to study, fairly and impartially, the city's environmental policy in its entirety and, if necessary, offer advice and recommendations.

- Finally, the ordinance specifically provides for a new system of citizen participation in the environmental policy, inviting citizens to express their views on the formulation of a Basic Environment Plan and on annual environment reports.

4 Future Prospects for Kawasaki's Environmental Policy

From now on, Kawasaki's environmental policy will be guided by the ideals and principles embodied in this Basic Ordinance, promoting various environmental measures that have been systematically incorporated into a Basic Environment Plan. The plan will be promoted in a comprehensive manner, incorporating environmental considerations in all policy measures and public projects by means of the environmental impact assessment system.

The Basic Environment Plan will thus serve as the guiding principle of the city's environmental policy. It will be adopted on the basis of a draft plan now being formulated, with coordination efforts taking place throughout city hall. Subsequently, it will be subject to a citizens' review and deliberation by the Environmental Policy Advisory Council.

The substance of the plan is expected to consist of the desired environmental situation that will serve as the objective of the environmental policy. It will identify targets for each environmental factor, specify basic environmental policy measures, and identify environmental considerations for the utilization of environmental resources.

In order to assure optimum management of the plan's implementation, annual reports will be prepared. These reports will be publicized each year, informing the public of current environmental conditions, the conditions of policy measures being adopted for the conservation and creation of the environment, and the state of environmental impact studies being conducted. Policy measures implemented in subsequent years will reflect the content of these reports, along with citizens' opinions and reports by the Environmental Policy Advisory Council on its deliberations.

Environmental impact studies constitute one of the major pillars of the environmental policy. Their purpose is to ensure that environmental considerations are fully integrated into the city's major measures and projects during the process of their formulation. These considerations will be a prerequisite to the adoption of all environmental policy measures—a safeguard that will guarantee the consistency and integrity of the city's policy measures regarding the environment.

With regard to citizen participation in environmental control regulation, a variety of forms can be conceived. These include the following: submitting opinions on the draft of the Basic Environment Plan, as provided for by the Basic Ordinance for the Environment; public hearings as stipulated in the Environmental Impact Studies Ordinance; and systems of public inspection of documents and citizens' written opinions, as stipulated by the Urban Planning Law.

Recently, however, citizens have been demanding to participate directly, from the formulation stage of the plan toward its development through direct discussions with city officials. This form of direct citizen participation is found in part in the development of park facilities, but it is quite uncommon in city administration as a whole.

The reason for this relative absence of direct citizen participation is that many citizens support the general principle of the plan, while often objecting to its particulars. This makes it very difficult for the city administration to coordinate the diverse opinions and demands of individuals within a limited amount of time. It is, however, a problem that will have to be addressed in the future by all concerned in municipal government.

5 Concluding Remarks

Based on its past experience with environmental policy centered on industrial pollution and in consideration of the nature of new environmental problems, the city of Kawasaki has promulgated the Basic Ordinance for the Environment. The ordinance is designed to implement a new horizontally-oriented, comprehensive environmental policy for the twenty-first century.

The general public's interest in environmental issues has been heightened recently, and people have become more environmentally conscious, in part because of the attention focused on global environmental problems. There remain, however, quite a few problems to be solved in the coming years for the new public policy regime to become well established under the current national systems of law and administration.

Institutional reforms are needed so that local governments can cope with local problems with greater autonomy. The necessary reforms include revisions to the present legal system, increased autonomy and greater fiscal independence of local governments, and reinforced cooperative relations among neighboring local bodies. Also called for is the reorganization of the city's internal administrative apparatus, so that appropriate environmental considerations may be incorporated in all policy measures and projects undertaken by the city—as well expanded citizen participation in the city's decision making process.

As new life-styles evolve, it will become necessary for the city, its residents and its business firms to become more aware of the environment and continually behave and engage in activities that are in harmony with environmental preservation.

A steadfast promotion of this new environment regulation policy will lead to sustainable economic development and contribute greatly to the creation of a desirable urban environment and the preservation of global environmental health. We will continue to make steady efforts to firmly establish this new system while attempting to resolve specific issues.

III.5 Remarks

Mr Umashashi

When we talk about sustainable development, there seems to be something that we may not readily see. Who are to make the decisions? Regarding this, I would like to talk on the role NGOs can play, since the participation of NGOs into the global policy making process has become increasingly significant. The significance and the advantages are as follows.

First, we see many adverse results in the developing nations and NGOs can reflect the voices of the people better. Secondly, NGOs do not have nationality, or are multinational. Some NGOs are active in a number of countries and they tend to be oriented toward the developed countries. Thirdly, when NGOs were involved in the early stages of policy making the coordination of opinions could be conducted smoothly. It meant there was enough time for coordination and adjustment. This is true for both the international level and the domestic level.

At the Rio Summit, the governments had to prepare the internal report with regard to their views on the environmental issues of the globe, and that meant that they had to have domestic committees. It was only natural, therefore, in Rio that there were many NGO representatives as part of the government delegations, especially in the Nordic nations. For example, there were 30 delegations from Sweden and of the 30 about seven were actually NGOs.

Now, there was one question I had about technology. It seems that it is almost a monopoly of the developed nations, especially with regard to environmental issues, and that these are being one-sidedly introduced by the developed nations to the developing nations. If that is the case, there are going to be some dilemmas or conflicts here.

Let me make my final comment. I am personally involved in the tropical rainforest issues. I have been in Borneo a couple of times and I felt then that we should think about this in a very natural way. People live in the tropical rainforest area and they have a sustainable life. But, through external pressure, they can no longer live in their accustomed ways. That is where the problem lies. That is the way I see it.

Chairman
Thank you very much.
Now some remarks from former Vice-Minister of the Environment Agency in Japan.

Mr Watanabe
Thank you Mr Chairman.
I resigned from my post at the end of June so I speak freely in my capacity as a private citizen. As once a person responsible for environmental policies, there are a number of thoughts that I can share with you.

The quality of the environmental issues has changed dramatically during the past 20 years. In the past, we talked about pollution control and health problems. But more recently, as is represented by global warming, we have come to realize that the issues are of a global nature and the domestic approach is not sufficient.

Speaking on Kawasaki, there we saw a different emphasis. People talked about the open type of lifestyle and pollution. Smoke stacks were the sources of pollutants in the past. Because of this, CO_2 restrictions cannot be achieved by companies alone, but rather we believe that the citizens, the people at large, can be both victims and polluters at the same time. I believe that such understanding of the concept is essential. The central government, the regional local autonomous bodies, the companies, businesses, NGOs and other institutions, and also each individual citizen, should seriously take up those issues as their own.

The basic laws concerning the pollution control as well as the basic laws for nature conservation have been the two pillars of the Japanese policies. But now we come to realize that new legislation of a more comprehensive nature will be required to incorporate such new concepts, because an administration can only act on whatever laws are in place. The basic law on environment has been drafted with the participation of all concerned agencies. Prior to that we tabled this issue before the experts in the Advisory Council. But laws are not enough. Some concrete actions are needed. I believe everyone must participate and the consumers' behavior should also change. But this is not an easy thing. One cannot force them to change their behavior. Rather, we should provide them with appropriate and correct information. Accurate information can enhance public awareness, and only with a heightened public awareness will people be persuaded to change their own behavior. For that, education is essential. Lifestyles should be changed, and for that we should be creative in devising new institutions inclusive of such behavioral changes.

III.6 Session Discussions

Dr Salim

First, how can Kawasaki arrange its environmental plan as a subsystem of the central government? How do you manage this balance? Second, how are we to reconcile the role of the Kawasaki Parliament on the one hand, and the citizens on the other? Which is to have the immediate impact on the decision making process?

Mr Kurihara

Up to now, the environmental policy of Kawasaki City is basically along the lines of the national policy, but maybe much more strict. This is specific to the city of Kawasaki because there were problems that Kawasaki itself had to solve on its own, for example the volume restriction that I mentioned earlier. This was set prior to the national policy. In the city of Kawasaki there is a location, roughly 10 percent of the city, where we need tighter environmental regulations and standards of our own. Regarding the policy making, Parliament of course is representative of the citizens. The citizens' participation is very important. We, the city officials, try to incorporate as many as possible of these opinions into our policy making as well as its implementation.

Dr Nakamura

If I understand correctly, Kawasaki has for a long time introduced a supplementary charge on SO_2 emissions in addition to the national scheme of the pollution related health standard compensation system. I just wonder what happened to this supplementary charge once the national scheme was amended in 1988. Secondly, we all live in a new economic recession. I wonder what is the direction of municipality regarding the question of how to preserve existing industry or to track a new industry. Would you be more in favour of compromises regarding a more strict environmental policy compared to former periods?

Mr Kurihara

If the Mayor was here, of course, he would be able to answer directly. We have a basic environment ordinance and the top priority is the relationship between development and environment. The basic idea behind this will not change. So we will not loosen the regulations.

Dr Cho

I have a question on Mr Kurihara's paper. Mr Kurihara, your explanation on Kawasaki's environmental plan is very informative and constructive, especially to a city which is planning a new environmental program. But you missed the explanation on how your city financed for the implementation of such a huge plan. Specifically, I want to know if there is any sharing between national government and local government.

Mr Ishida

I have a question to Mr Kurihara too. You talked that the citizens would accept the framework or the general concept, but when it comes to individual concrete interests involved, they tend to oppose. Kawasaki is really advanced in implementing the environmental policies. I guess the citizens may have changed their own lifestyle or behavior themselves. Was that type of change observed?

(Interpreter)

I would like to ask Dr Sprenger a question. I think that Germany is taking the initiative right now, but what concerns me is that if you try to actually implement projects as you said in your remark at the end, then cross-boundary problems, such as EEC issues, may be used as an excuse and the project will not be realized. There are often conflicts of interest within and among countries. Is there some reasonable way to solve this problem?

Dr Sprenger

Before replying I would also like to raise a question with Mr Watanabe regarding the Japanese approach. As I understand, Japan is on its way to designing a new basic law for environmental protection and there are some ideas or initiatives to include in this new basic law—more options regarding the use of economic incentives and instruments. This raised a question from my perspective, whether you are more critical nowadays of the mix of instruments you have used in the past, including Kagaibashi, Kyotei? My specific comment is that Japan has to look for less expensive solutions to comply with existing regulations in order to preserve the existing environmental standards.

Chairman

I would like to raise a question with Dr Suraphol Sudara. You talked on the interaction between NGOs and international environmental protection agencies, but it seems to me that we have missed your evaluation of what your government is doing. How do you view the function of government, both within the country and globally, in finding a solution to the need for environment protection.

Dr Nakamura

One question with respect to Dr Sprenger's paper about the pollution charges on waste water. You mentioned that they have been amended three times. Could you elaborate a little more on the process of revision and the impact it had?

My second question is for Dr Sudara. Mr Watanabe was mentioning that it is getting to the point where we are, on the one hand, the victims of pollution, but at the same time we are generators of pollution and we have to take responsibility. The Kawasaki case was that a more sophisticated process of citizen involvement in the decision making was one way of addressing this, while you are saying, in the case of Thailand, that there is still quite a lot of conflict between the development oriented government and the people in general, particularly in the areas where you do not get the benefit of economic development. How do you solve this polarization?

Dr Sudara

I would like to respond to both questions at the same time. Actually we have been trying to work with the government, but at this stage most of the governments in developing countries are still not quite sure which way they should take regarding the people participation, and getting the information to the people. So what we are doing is that many of us are sitting in on various committees so that we can input our ideas at the beginning. We would also like to have all the information open to the people. Once the information is open, then people can have the reaction from the beginning. This approach has proved to be successful—for example, with the dumping of toxic waste. There are many other issues that helped raise the consciousness of the indigenous peoples. Some groups have been recognized by the government as partners in the environmental policy making process. Many more will follow.

Dr Sprenger

I start with a question of Dr Nakamura regarding the waste water charge law. The amendments were mainly dealing with the introduction of new parameters, especially for the heavy metals. Then we have changed tariffs. We changed the conditions for exemption because in many cases the investment for new waste water treatment plants or new technologies could be compensated against the charges, and so the general impression with respect to the last amendment is that the waste water charge law has somehow become some sort of economic incentive. It has degenerated more or less to an instrument of subsidies for the industry and so is breaking the polluter pays principle. The pioneering activity of Germany disappeared, so that country cannot be regarded any more as a model for other countries.

Regarding Professor Ishitani's question whether easy legislation is often used as some sort of excuse for not doing more or for not progressing on the national level, I fully agree. I wonder whether the approach you have mentioned, to use ISO standards as an approach to improve the environmental situation with respect to products, might be one which is generating progress. I feel that the influence of industry is very strong and looking at the present economic climate I can't see why they should have a strong interest in fixing standards that are more strict.

I feel also that the international standard organizations could be just another very subtle means by which the industrialized countries prevent other countries coming into the markets. We from the industrialized countries cannot regard ourselves as a model. We are still depleting the natural resources in the Third World in an increasing way that is posing a lot of risk to our planet. We are dictating prices, so it is not a question of having fair trade and giving financial resources to these countries to develop in a way which we would call today sustainable development. At the same time, we are building up new barriers for these countries in terms of tariffs on products that pose less environmental risks, and fixing non-tariff barriers by requiring more stringent environmental standards for most of these countries which they cannot comply with partly because we have marketed products like pesticides in these countries which we don't want to have in our products or in our imports.

We are not giving good example regarding lifestyle, the consumption pattern, or the volume of consumption. Look at, for example, the increasing consumption of gasoline in the western world. We are trying to compete with the less industrialized and developing countries with low cost by sacrificing our own environment or by ignoring our own standards. In many cases, for example, as I have mentioned, even Germany and the EEC countries are not implementing the EEC standards. We are requiring the standards to be respected by the developing countries while we ignore them ourselves. In the race for new markets and for employment we are sacrificing the environment. It's a race where neither the environment nor the employment side will win.

Mr Kurihara (Interpreter)

Due to time I shall be very brief.

Concerning finance, we do not have the firm financial arrangements to implement the policies needed. Under 11 pillars, there are about 47 concrete environmental policies. Some of these on their own would require trillions of yen, maybe, so we are talking about a large investment. But we believe that the drafting of such a basic principle has been significant in the sense that we can use this to persuade the authorities to allocate more funds to such efforts. But we do not have the funds in place to implement all those. That will have to happen in the future.

The next question was about whether there has been a change in the lifestyle or consumption behavior of citizens. Again, the ordinances and the basic plan alone do not warrant the immediate change of lifestyle. Citizens may react to critical issues such as waste or garbage disposal because they realize that there are no places to fill or dump those wastes. But in other areas where they do not have a direct impact, it will take time to change their behavioral patterns. So we believe that more and more enlightenment will be needed to induce a change in the pattern.

You asked for an example of the participation of citizens. When roads are being built, people often support that because it will enhance their convenience.

But when that road is located next door to their house, they oppose it. So they may agree on the concept but do not necessarily support the implementation of concrete plans. That is a reality that we face.

Mr Watanabe
Regarding the new bill, the new basic law does provide economic incentives. It calls for payment of tax credits and economic incentive subsidies. Also, there are other incentives to discourage people to be engaged in actions which would be detrimental to environment. The conventional instruments were helpful in overcoming the conventional scenario of confrontation between the victims and the polluters. But the new law has been drafted not to rectify the shortcomings but to try to accommodate the needs of the present time because both the issues and people's natures have changed. We are trying to induce behavioral changes in consumption, so the bill is designed to introduce not only positive incentives but other types of incentives which encourage people to change their consumption pattern.

Chapter Four

International Collaboration toward a Global Environmental System

IV.1 Overview

This chapter includes three papers which discuss international collaboration toward a global environmental system. Of course, as is common, the three papers take different views to the areas that need international collaboration.

The first by S. Kato explains what the Japanese government has so far done in order to promote Japan's collaboration in international activities. This brings us to the domestic problems and efforts in tackling the contradictory demands at home, namely the continuation of economic development on one hand and the enhancement of environmental protection on the other. Because of the considerable size of Japan's economy and its industries, the protection of global environment is in its own interest as well. Hence, he argues, it is the utmost task of the government to lead their country's whole economy toward being sustainable and interdependent and to do this by transforming the way of thinking in Japan, viz 'development in one country'.

The second paper of Y. Cho also stresses the importance of international collaboration not only from the Korean perspective but also from the regional perspective of Asia and the Pacific. In this regard, he argues, there emerges a need for incorporating various approaches in various countries into a regional process, perhaps through APEC (Asian Pacific Economic Cooperation). With such an arrangement at regional level, the protection of global environment can be more meaningfully implemented.

In the last paper, N. Nakamura brings us to the real state of the situation. He introduces the problems faced by those living around an inland lake, Biwa, and the efforts made to solve them. Though useful as an experience, more important is the fact that those concrete problems are not isolated phenomena unique to Lake Biwa. On the contrary, as he ably explains, they are part of the global problem. From this, he proposes the establishment of an international cooperative network for a specific issue, rather than for all issues. This is another important point of interest to be looked into more seriously.

IV.2 International Collaboration: the Role of Japan

S. Kato

President, Research Institute for Environment and Society, Japan

Abstract

The Earth Summit, which took place in June 1992 in Rio de Janiero, Brazil, succeeded in forming 'global partnership' and agreements on a number of important and urgent issues such as climate change, the protection of biodiversity and forest management.

However, a little over one year has now passed since the Earth Summit, and the results have fallen sort of the expectations of the world community. Commitments made by the industrialized countries are yet to be fulfilled, and the devastation of the environment in many parts of the world has not stopped.

Japan has contributed, and will continue to contribute, to strengthening the global environment in a number of ways. Firstly, after the Second World War the Japanese government enforced strict environmental regulations, and transformed itself into an energy and resource saving society, with support from the business community. Secondly, Japan supports the efforts of other countries, especially developing countries, with environment-related overseas development aid.

The Japanese government has drafted a new basic law for global partnership and environment protection. It is hoped that this will become the basis for Japan's contribution to protecting the global environment.

1 One Year after the Rio Earth Summit

It has been one year since the convening of the United Nations Conference on Environment and Development (UNCED) or Earth Summit which was the highest-level and largest conference organized by the United Nations, joined by about 180 governments and attended by the heads or sovereigns of more than 100 states. The Summit was a historic achievement because it established the world's concern about environment and development and produced consensus on global environmental protection at the highest government level. Without doubt it was a success. It is now necessary to maintain the momentum of the Rio Summit and follow up the outcomes of UNCED and to promote continuing efforts for environmental protection.

I would like to look at the movements of the world since the Summit. The United States is changing its policies in favour of the global environment, having signed the Convention on Biological Diversity, and is strengthening its policies to tackle global warming. There will be a stronger possibility of using collaborative efforts to tackle global environmental problems among developed countries. Many developing countries are trying to achieve sustainable development, pursuing national socio-economic growth while safeguarding the environment, but this is not an easy task for them because of limited financial resources and technological capacity.

The Commission on Sustainable Development (CSD) was established under the Economic and Social Council of the United Nations, based on the agreement of Agenda 21. Its first meeting was organized from June 14 to 25, and to ensure effective follow-up meetings of UNCED and CSD are planned to be held every year.

Efforts to restructure the Global Environment Facility (GEF) (which was established to assist developing countries to protect the global environment and to transfer environmentally benign technology, as well as serving as an interim financial mechanism for the climate and bio-diversity conventions are underway. These will examine expanding its scope and coverage to relevant program areas of Agenda 21 with global environmental benefits and ensure a transparent and democratic governance.

The current efforts, however, in both industrialized and developing nations to take effective action are far from satisfactory. The commitments made by industrialized countries are yet to be fulfilled, because of social, political and financial difficulties. The devastation of the environment in many parts of the world has not stopped. On the contrary, it is actively underway, causing disasters in a number of areas, including Japan.

2 Japanese Experiences and Lessons

Twenty-one years ago, when the United Nations Conference on Human Environment was held, Japan had high economic growth and suffered from serious environmental problems. Health damages caused by pollution, such as Minamata disease by mercury poisoning and Yokkaichi asthma by air pollution, water pollution in rivers, lakes and seas, and the degradation of natural environments drew public attention both here in Japan and abroad. Under these circumstances, an extraordinary session of the diet which latter became known as the 'pollution diet' of 1970 was convened and many environmental laws were enacted. The Environment Agency was established to carry out the planning, coordination and implementation of environmental policies and measures.

However, the road to environmental protection has been far from easy or smooth. Thanks to the determined efforts of all people concerned, such as community residents, local governments and private enterprises, some of the serious environmental problems have been overcome. We have developed many environmental technologies such as pollution reduction, monitoring and process innovation which set precedents to the rest of the world.

Japan has learned many important lessons in this process. Environmental pollution not only caused tragic health damages, but was accompanied by the costs of compensation for the damages or of restoration of the degraded environment. These costs proved much greater than those of implementing measures that would have prevented pollution from occurring in the first place. These experiences in Japan tell us that conducting economic activities without adequate consideration to the environment does not pay, even from a strictly economic point of view. I do hope that every country may learn from these lessons and achieve environmentally sound economic development, taking into account the importance of prevention rather than cure.

We have been able to reduce industrial pollution through the development and application of stringent emission control standards and pollution abatement technologies. As a result, Japan finally managed to transform itself into an energy and resources saving society and dramatically improved its environment. Thus the Japanese experience so far has shown to the world that environment and development are not only compatible, but also mutually reinforcing in the long term. I believe this point is very important.

'Japan has established a remarkable record in combining substantial improvements in its domestic environment with unprecedented levels of economic growth. This provides the basis for Japanese leadership in global environment and development cooperation commensurate with its growing weight in the world economy. We appeal to Japan, along with other industrial nations, to assert their leadership at the Earth Summit.' (Tokyo Declaration adopted by 'Eminent Persons' Meeting on Financing Global Environment and Development', April 17, 1992)

'Japan, thanks largely to its own efforts, has transformed itself from a war ravaged developing country to an economic superpower. This extraordinary achievement is universally acknowledged, widely admired and occasionally envied. Much less known or appreciated than its economic miracle is Japan's second miracle—the remarkable progress it has made in the past two decades in cleaning up its domestic environment. With an astute combination of tough regulations and liberal incentives, Japan has succeeded in reducing domestic air and water pollution levels and achieving efficiency in energy and raw materials use more than any other country. And it did this while maintaining the most dynamic economic performance of any industrialized nation. Japan's experience has demonstrated convincingly that good economic management and environmental protection go together, that economic efficiency is the best route to environmental sustainability. In this course of Japanese policy makers and business leaders have established a strong competitive advantage in pollution abatement technologies and industrial processes.' (Maurice F. Strong at the opening ceremony of the above mentioned Eminent Persons' Meeting, April 15, 1992)

Yet even today, we face a variety of difficult problems. For instance, the ever growing scale of socio-economic activities delays the solution of problems or brings new problems to our society. For example, although we set the strictest emission standards for motor vehicles in the world, the number of automobiles, in particular that of diesel engine cars, has been growing so rapidly that still today the general air quality in big urban areas like Tokyo and Osaka leaves much to be desired. Another example is that because of increasing diversification in our life-styles, both the quantity of wastes generated by household and industry and the kinds of hazardous substances contained in them are increasing.

To a large extent, this may be attributed to the socio-economic structure of developed countries which are dependent on mass production, mass consumption and mass wastage. Japan needs to change its socio-economic structure to an environmentally sound and sustainable one, while ensuring stable economic growth. This approach is particularly relevant in dealing with global environmental issues, such as curbing CO_2 emissions.

3 The Role Japan Should Play

Japan maintains large-scale economic activities and depends on imported resources, so it is closely linked to the health of the global environment. At the same time, Japan possesses a wealth of experience and technology as well as economic power. I believe that Japan's primary role to the global environment is to take a lead in reforming our society to an environmentally sound and sustainable one. The second responsibility of Japan is to extend technical and financial assistance to developing countries and to contribute to forming a workable consensus among developed and developing countries.

I am confident that the Japanese people and government recognize that the most suitable contribution Japan can make to the world community would be to attach the highest priority to helping developing countries address global environmental problems.

4 Contributions Japan Should Make

4.1 Japan in the Asia-Pacific region

Needless to say, Japan is situated in the Asia-Pacific region. The region spans a vast and diverse range of eco-systems including deserts, forests, rivers, lakes and seas; a large part of the region is marine. The marine and coastal environments of the region are among the most diverse and productive in the world. The climate in the region also extends from the poles to the tropics. With only 23% of the world's total land area, the region is home to 55% of the world's population. Many megalopolises have emerged as a result of the population shift toward the cities.

With its significant human and natural resources, the region's economy is expected to become the world's most dynamic as the twenty-first century approaches. Consequently, interdependence among countries will be strengthened and the flow of people and commodities will also be expanded.

On the other hand, economic growth is expected to spur the use of energy and natural resources. Without proper policies and management, air pollutants such as SOx , NOx and CO_2, water pollutants such as organic compounds and heavy metals, as well as hazardous wastes would increase sharply, while natural resources including forests, land, fresh water and biological diversity would deteriorate. Such environmental degradation could have impacts even beyond national borders and regional boundaries, extending to a global level. From the viewpoint of the regional impacts of global environmental problems, fragile eco-systems such as mountains, lowlands and small islands, which can often be found in nations of this region, could be damaged.

The Asia-Pacific region is in a crucial era in terms of steering development in a sustainable direction and pursuing socio-economic development while safeguarding the environment. Therefore, it is imperative for each nation to strengthen its own efforts to achieve these goals, as well as to promote regional cooperation and establish new environmental partnerships to tackle common problems. Such endeavours in the region could act as models for other regions, especially developing regions. The Japanese initiatives in promoting such actions in Asia and the Pacific are believed to be increasingly important.

4.2 Necessity to have solid long-term perspectives for the region

It is necessary to have a long-term view, beyond the present generation, in order to satisfactorily cope with the issues of environment and development as well as to achieve sustainable development. Accurate and adequate information, not only on the present state of the environment but also on trends and future prospects for environment and development, can provide decision makers with a scientific basis for policy formulation. Drawing up such perspectives would also contribute to the establishment of shared perceptions among policy makers in the region, and to the promotion of regional cooperation. Participants to the ECO ASIA '93, which took place in Chiba, Japan in July this year, with participation of senior officials from 16 governments, including nine ministers, plus representatives of seven international organizations, welcomed the initiatives of the Environment Agency of Japan on its new project 'The Long-term Perspective on Environment and Development in Asia and Pacific'. It is hoped that this project will be conducted with the active cooperation of other governments in the region, relevant international organizations and research institutes.

4.3 Japan as a technology center

Through JICA's technical assistance programme, Japan has provided a variety of technical assistance to the Third World countries for many years. These include training personnel in Japan, sending Japanese experts to developing countries and setting up research and training centers etc in these countries. I am quite sure that Japan will continue to extend various forms of assistance to developing countries to address problems of pollution, waste management, deforestation, desertification and soil erosion. This assistance will aim to realize sustainable development in developing countries. Japan will also promote regional cooperation in diverse areas such as global warming and bio-diversity in the Asia-Pacific region, and acid rain, marine pollution and migratory bird protection in North-East Asia.

Last autumn the Japanese government invited UNEP to open its International Environmental Technology Center in Japan. The Center's main objective is to promote the transfer of environmentally sound technologies and know-how to developing countries. The government will support its activities, together with the hosting local governments. Creating a network among this UNEP Center and other national centers of excellence would be an idea worth exploring in the future, including networking with environmental research and training centers of Thailand, China, Indonesia and others, to which the Japanese government is extending financial and technical assistance.

4.4 Japan as a financial supporter

Japan has supported the efforts of other countries, especially developing countries, through existing bilateral and multilateral mechanisms. Self-help efforts on the part of developing countries are of primary importance to make such support truly effective. In fact, Japan announced, at the Arche Summit in 1989, its target of committing around Y300 billion (US$2.3 billion) of environment-related aid during the three fiscal years 1989-1991. Actually, we have exceeded that goal by providing more than Y400 billion (US$3.1 billion) within that time frame. Japan pledged at the Rio Conference to expand its bilateral and multilateral overseas development aid in the field of the environment to around Y900 billion to Y1 trillion (US$7 to 7.7 billion) during the five year period starting from fiscal year 1992. Partnership with developing countries is indispensable for the successful implementation of environment-related overseas development aid. Japan should, therefore, do its utmost to find, formulate and implement effective projects through consultations with developing countries. As the Tokyo Declaration of the Eminent Persons' Meeting stated, financing for sustainable development should not be viewed as 'foreign aid' in traditional terms. It is an essential investment in global environmental security. It calls for forging a global partnership for developed and developing countries to cooperate for global interests.

IV.3 North-East Asian Collaboration: Toward a Global Environmental System

Y. Cho

Professor, Institute of Foreign Affairs and National Security, Seoul, Korea

Abstract

Last year's United Nations Conference on Environment and Development (UNCED) in Rio de Janiero has illustrated the vital importance of environmental protection. Environmental destruction, such as the depletion of the ozone layer, global warming, acid rain and marine pollution, has now become a global as well as a regional problem.

Despite the urgent need for cooperation to deal with these problems in the North-East Asian area, it was not until recently that collaborative activities started among the countries of this region.

The countries of the North-East Asian area show great diversity. For instance, the region spans, geographically, from arctic to semi-tropical zones. It includes countries with substantially different political and economic systems, and also includes various levels of economic development, from developing and newly industrialized countries to highly developed economies. In addition, regional environmental problems which call for urgent action range from air pollution to marine pollution. In short the environment of this region can be regarded as a microcosm of the global environment. Thus, concerted and solid efforts from these countries are essential to deal with the problems.

The current status of cooperative activities for environmental protection in the region remains at a disappointing level. There have so far been a few attempts to set up regional frameworks to address outstanding environmental issues, yet the results have not been encouraging. Although a permanent mechanism for environmental cooperation has yet to be established, the efforts to hold regular inter-governmental conferences have been stepped up.

The starting point for solving regional environmental problems should be to set up a mechanism to overcome the inequality of environmental information between countries. The next step would be to conduct joint research and find optimal policies in economic and technological terms. The financial difficulties in executing related projects can be addressed as the last step.

Present discussion on environmental cooperation in this region tends to focus on securing financial sources and sharing the burden. Given the conflict between the economic interests of the countries involved and without sufficient information on the current status of pollution, it will make the process slower. Therefore, securing the relevant and comparable data on environmental pollution of the region should be given the first priority in the discussion of existing regular inter-governmental conferences, which would make it easy to proceed to the next step for establishing proper cooperative policy.

1 Introduction

Last year's United Nations Conference on Environment and Development (UNCED) in Rio de Janiero has illustrated the vital importance of environmental protection. As environmental degradation goes beyond nature's self-purification capacity, nature itself is putting restrictions on sustainable development, and threatening the subsistence of human beings. Environmental destruction such as the depletion of the ozone layer, global warming, acid rain and marine pollution have now become global problems as well as regional problems.

The North-East Asia region comprising China, the Democratic People's Republic of Korea, Japan, Mongolia, the Republic of Korea and the Russian Federation is not an exception. This region boasts the most dynamic and productive industrial area in Asia. Many countries of this region have given priority to the development of their economies at the expense of the preservation of the environment. The rate of economic growth in the 1980s in most of the region has been remarkably high, with China and the Republic of Korea registering rates of growth of 9.5% and 8% respectively. The major emphasis of the countries' policies has been on industrialization. The results are that the industrial sector's share of gross national product (GNP) has improved in all the countries during the 1980s, and that the consumption of energy, including coal, petroleum and nuclear energy, has increased very rapidly. Urbanization has also proceeded considerably as a result of a concentration of economic activities in a few sites.

As a result, the environment in this region has suffered and has been deteriorating at a rapid pace while the rate of economic growth has been tremendous. For this reason, since the beginning of the 1990s, environmental issues have become important economic, political and diplomatic concerns. Though each country in this region has made an effort independently to curtail the emission of pollutants, the countries have other problems which they cannot manage by themselves. The nature of the environment of North-East Asia is basically enclosed, so the major environmental problems would be trans-boundary air pollution and marine pollution of the Yellow Sea and the East Sea (or the Sea of Japan), which are basically regional and require cooperation among the countries for their solution.

Therefore, regional cooperation is essential for the region's environmental protection. Until recently, however, such cooperation has not been active, which was partly because the socio-economic conditions of the countries in this region were quite diverse. Now, the elimination of the cold war confrontation has given these countries the chance of closer cooperation for environmental preservation. Several programs are being activated to support regional cooperation.

The purpose of this paper is to survey the current state of environmental pollution and ongoing regional cooperation, and to try to make proposals for a desirable regional environmental cooperation which takes into account the complex environmental conditions of this region.

2 Current Environmental State in the North-East Asian Region

2.1 Current environmental conditions

The North-East Asian region shows great diversity in its approach to socio-economic development as well as in its histories, cultures and religions. This region includes the free market economies of Japan and the Republic of Korea on the one hand and socialist economies of China and the Democratic People's Republic of Korea on the other. There are also the economies of the Russian Federation and Mongolia which are presently in transition toward the free market system.

Despite such diversity, most countries of the region have experienced a remarkably high rate of economic growth during the 1980s, which is accompanied by rapid industrialization, and requires exploration and exploitation of new and greater sources of energy for the support of economic development.

China ranks first in the world in its output of coal and cement, and has become a major country in the production of steel, crude oil and power generation. The strategy for industrialization in the Democratic People's Republic of Korea entails increasing the capacity to produce fertilizers, agricultural chemicals and cement, and expanding heavy industries. The Russian Federation has also developed considerably in heavy industries including the production of armaments. The Republic of Korea is leading newly industrialized countries, and the industrialization has caused rapid urbanization and related environmental problems.

As the harmonization between development and environment was ignored in its early stage, such rapid industrialization has contributed much to the environmental degradation of this region. Furthermore, the rapid increase of energy consumption accompanying the industrialization is aggravating the situation.

The primary source of energy both in China and in Mongolia is coal. It accounts for 75% of the industrial fuel and power, and 85% of fuel for urban residence in China, while in Mongolia it accounts for 90% of all domestic sources of energy. About 60% of Mongolia's total coal consumption is used by thermal power plants, and the quality of the coal is very poor as it has a high sulfur content. Anthracite coal is a major source of energy, along with hydro-electricity, in the Democratic People's Republic of Korea, but its quality is very high with a low sulfur content. In the Republic of Korea, half of the energy consumption is attributed to imported petroleum, while 30% comes from coal and 15% from nuclear energy. But there has been a gradual shift from the use of coal to the use of petroleum. Increased use of cleaner energy sources such as LNG, low sulfur fuel and unleaded gasoline is being encouraged, and the development and dissemination of clean energy technology is being promoted. In the Russian Federation, thermal power dominates the overall energy production. Efforts to replace coal by natural gas in the thermal power plants has decreased the total emission of sulfur gases by 32% over the last 20 years.

2.2 State of environmental pollution

Because of rapid industrialization and the accompanying socio-economic development, environmental problems have been taken seriously in many urban and industrial sites.

Air pollution is recognized as one of the major problems of the region primarily caused by energy and industrial sources. Air pollution in China's sites is mainly caused by the smoke produced in the combustion of coal. Backward technology of the industrial sector and low energy efficiency result in high consumption of energy and natural resources, and thereby increase the level of pollution. The level of air pollution in the Russian Federation is very high in many areas, registering 10 times higher than the international standards in 84 towns and cities in 1991. Sulfur dioxide discharge is a main source of air pollution, and is common for all countries of this region. It is posing serious problems in China, the Democratic People's Republic of Korea and in the Russian Federation, while the concentration of sulfur dioxide in Japan and Korea came down from .06 ppm in 1965 to .01 ppm in 1990 and from .094 ppm in the 1980s to .051 ppm in the 1990s respectively. Although, in some countries, efforts are being made either to undertake desulfurization of coal and oil, or to shift the use of coal to the use of low sulfur fuel, the concentration of sulfur dioxide is still much above the permissible standard in other countries.

Another major concern expressed by all countries in the region is the increasing acid rain, which results from increased air pollution. Acid rain was detected in certain areas of north-west, central-south and south China with an increase of 8.7% in the number of affected cities from 1990 to 1991.

Water pollution also poses a very serious problem. While sources of water pollution differ among different countries, industrial waste discharges are a principal source. More than 70% of the total waste water discharges in China came from industrial waste in 1989, and 80% of them went into rivers, lakes and the sea without treatment. In the Republic of Korea, industrial waste water has been increasing at the rate of 20% per year, so more attention is needed as the quality of water deteriorates. Domestic sewage is also an important contaminant source both in Mongolia and in the Republic of Korea. In the Russian Federation, almost all water bodies are badly polluted, including the Amur River which flows into the Pacific ocean. Many enclosed water bodies in the countries of this region are polluted by nutrient accumulation as well as industrial discharges. As a result, eutrophication is an important shared concern.

Because of the discharge of industrial wastes and the run off of rivers in the countries of this region, the marine environment is also polluted in the Yellow Sea and the East Sea, which may require imminent concerted efforts from these countries for improvement.

2.3 Regional environmental concerns

In many cases, pollution is not confined within a country's borders. Since the late 1970s, environmental damage has often been caused by pollutants moving between countries. The diffusion of pollutants may result in international and inter-regional frictions which force the world to realize the necessity for concerted efforts to come up with a global or regional scale solution.

The countries in the North-East Asian region are close neighbors geographically and climatically, and most have common borders and shared sea waters as well as shared air. Hence, there have been some problems in this region which have affected a group of countries, such as the across the border movement of air pollutants and marine pollution. Thus, trans-boundary air pollution and the pollution of the Yellow Sea and the East Sea can be regarded as the two major environmental problems in the North-East Asian region.

Under the geographical and climatic conditions of this region, the Republic of Korea is vulnerable to trans-boundary movement of pollutants. The seasonal winds, originating from the Mongolian Heights, blow south-east via China to the Korean Peninsula and the Japan Islands. It means that the infamous yellow sands, as well as various air pollutants generated by the Chinese, are most likely to drift to Korea and Japan. This is very frustrating for Korea, which invests a great amount of its resources and efforts to cut down its domestic sources of emission.

The marine pollution of the Yellow Sea is becoming more serious, as most of the large rivers of Korea and China, flowing into the Yellow Sea, are heavily polluted. Moreover, China has already built huge industrial complexes on its Yellow Sea coast, and Korea is also trying to expand its production facilities on the other shore of the same sea in an effort to prepare for growing trade with China. Also, as the far eastern part of Russia has enormous potential for economic development, the East Sea is facing marine pollution from land based sources as well as shipping operations. Thus, this region is witnessing increasing pollution, and this is where the necessity for regional cooperation arises.

3 History of Environmental Cooperation in North-East Asia

Intensive effort has seldom been made for environmental protection in this region until recently. In general, it would not be easy to reach an agreement for international cooperation in environmental problems because a super-national enforcing entity is not often available. Even in cases where an agreement can be reached among the relevant countries, it does not tend to reflect a global optimal state under the given situation. This is due to the fact that global benefits (or costs) would not necessarily coincide with the sum of individual country's benefits (or costs). For example, when a certain trans-boundary pollution is uni-directional, the polluter country does not have as strong an incentive as the victim country to monitor and reduce the pollution. With this inequality of incentives, investment by the polluter country for reducing pollution cannot be executed up to the level of global optimum. In addition, establishing a regional agreement at the formal level will have another difficulty which comes from the wide diversity of this region. The region includes developing countries, newly industrialized countries and an economic superpower. Per capita GNP ranges from a few hundred dollars to more than $20,000. With respect to environmental management, Japan is already very successful, and is becoming a leading country in many areas. Korea is now starting active environmental management on a large scale. China and the Russian Federation have recently been recognizing more need for environmental protection, but still they seem to put priorities on their own development. China and the Russian Federation may need more funds and advanced environmental technologies, while Japan is more concerned with trans-boundary pollution rather than domestic sources of pollution. The different concerns of various countries of this region are making any solid progress in regional cooperation in environmental protection very difficult.

Despite these difficulties, some progress has already been made which supports regional cooperation in North-East Asia.

3.1 The North-West Pacific Action Plan (NOWPAP)

The North-West Pacific area is virtually the only regional sea which does not have a cooperative regime for environmental protection. In 1991, the UNEP initiated an activity for the preservation of the marine and coastal environment of the North-West Pacific as one of the regional seas programmes. Since the first meeting of experts and national focal points, which was held in Vladivostok, October 1991, the state of the marine environment of this area was reviewed, draft action plans were discussed, and the activities for the action plan were proposed. It was agreed that NOWPAP should cover the marine environment and coastal areas of the Yellow Sea and the East Sea. The various action plans that have been adopted or are still being developed, have a number of common objectives. The main objectives among them are to assess the regional marine environmental condition by coordinating and integrating monitoring and data gathering systems on a regional basis, to collate environmental data and information to form a comprehensive database and information management system, and to develop and establish a regional framework towards the integrated planning and management of the marine environment.

NOWPAP will be adopted by the inter-governmental meeting once the draft has reached the stage at which the adoption can be recommended to the governments of the countries of this region. After this, a decision making body for the action plan must be made up of senior representatives of the member countries. In the next meeting, which is scheduled to be held at Bangkok, November 1993, NOWPAP countries will be invited to set up a specific inter-governmental forum to provide policy guidance and decision making for the action plan. But, until a regional body is established, UNEP will play a role as the organization responsible for the coordination of the implementation of the action plan.

3.2 Informal frameworks of environmental cooperation

Marine and coastal pollution is only one of the many environmental issues in North-East Asia. Trans-boundary air pollution and radioactive pollution need more attention. An inter-governmental regional programme should be established for an efficient information system and for the coordination and implementation of various projects.

There are two significant developments in this regard. The Seoul Symposium and the Niigata Conference provided opportunities for the discussion of a comprehensive regional environmental management system of this region. Various issues were examined in the meetings, such as air and water pollution, urbanization, industrialization, energy use and eco-system degradation.

In the Seoul Symposium on UNCED and Prospects for an Environmental Management System in the twenty-first century (September 2-5 1992), high ranking officials of environmental agencies of China, far eastern Russia and Mongolia in addition to Japan and Korea were invited to attend for the first time. The participants of the countries agreed that they cooperate more closely with each other in regards to the environment, and that an informal network for environmental cooperation should be established. The network would promote the exchange of information and experts, and cooperative research and surveys related to the environment among the countries, and thereby promote environmentally sound and sustainable economic development within the region.

The North-East Asian Conference on Environmental Cooperation at Niigata (October 13-16 1992) was another informal meeting with the participation of China, Mongolia, the Russian Federation, Japan and the Republic of Korea. In the conference, the participants exchanged their opinions and information on the environmental state of each country and proposed various ways of promoting regional cooperation. Special emphasis was placed on air and water quality, marine pollution, waste management and recycling, bio-diversity and nature conservation. The participants stressed the importance of exchanging information, experiences and expertise on a regular basis, and also emphasized the necessity for a continuous dialogue for their common concerns. It was also agreed that the participants of various non-governmental organizations as well as public sectors would be important for efficient cooperation in this region. The next conference will be held in Seoul, Korea in September 1993.

The two frameworks mentioned above have contributed much to initiate the discussion of the development of a comprehensive regional environmental system. They succeeded in exchanging opinions and information, and identifying several common concerns, such as conducting joint surveys and monitoring activities on environmental pollution. They also succeeded in carrying out collaborative research activities in some areas and establishing the network for exchanging information. However, they still remained at an informal level and have their own limitations in the sense that they do not adopt any enforcing resolution or regional agreement among the countries.

3.3 The North-East Asian Regional Environmental Programme (NEAREP)

In the keynote speech in General Assembly of ESCAP, April 1992, the Minister of Foreign Affairs, Sang-Ok Lee of Korea, proposed the implementation of a research project on the establishment of the North-East Asian Environmental Cooperation Programme. Two months later, during the UNCED held in Rio, Prime Minister Chung of Korea announced that his country has a genuine interest in establishing some kind of organization to oversee environmental management in the North-East Asian region. In the fifth annual conference of Korea-ESCAP Cooperation Fund (KECF) at Bangkok, July 1992, it was agreed that the North-East Asian environmental meeting supported by KECF should be held. Following this procedure, the Senior Officials Meeting on Environmental Cooperation in North-East Asia was held, February 8-11 1993, at Seoul, Korea. The five countries of this region participated under the aegis of ESCAP and in cooperation with the UNEP, UNDP and the ADB.

The participants of the meeting expressed a strong desire for promoting regional cooperation in North-East Asia, and recognized the necessity for a sustainable inter-governmental meeting for regional environmental problems. It was also agreed that while the experiences of regional cooperations in other areas will serve as a good reference, the North-East Asian regional cooperation should meet the specific requirement of the region reflecting the region's diversity in socio-economic conditions. That is, the cooperation in this region should start from the cooperation of priority areas at its initial stage, and gradually proceed to the establishment of an overall strategy and its implementation planning.

Thus, most of the countries enumerated various areas of concern in which they would like regional cooperation to be promoted. Among the many areas of concern raised in the meeting, the following three areas of concern were identified as priority areas within which specific projects for regional cooperation should be considered and developed at the next meeting scheduled for Beijing, China: (a) Energy and Air Pollution; (b) Eco-system Management (in particular, Deforestation and Desertification); (c) Capacity Building.

The coastal/marine pollution problem was cited as another priority issue and it was felt that this concern could be covered within the framework of UNEP/NOWPAP.

Air pollution is considered to be the most important environmental issue, as the remarkable rate of growth achieved in the countries of the region depends mainly on the availability of sources of energy such as coal and oil, (as described in the previous section).

The major tasks to be discussed at the next meeting are to re-define priorities for collective and individual actions and to provide financial, technical, managerial and human resources to implement them in an effective manner. Thus, capacity building efforts on a regional scale should be important to effectively encourage cooperation. Capacity building includes activities such as sharing and exchanging information, training and human resources development, and technology cooperation. Fortunately, the region has a few countries which have developed highly sophisticated environmental technologies, and many countries have developed advanced systems of pollution control and waste management. Therefore, joint endeavor in capacity building could assist the developing countries of the region to launch substantial programs of environmental management.

For further and regular meetings of the senior officials in the future, ESCAP is asked to provide administrative and financial support in cooperation with UNEP, UNDP and the ADB until the institutional arrangement of the meeting is established.

In connection with the organization of the next Senior Officials Meeting, the ESCAP secretariat, through consulting with the countries, is developing program proposals with identified activities in the priority areas for presentation at the next meeting. The meeting at Seoul recommended that the UNDP provide necessary financial support for this purpose, which is available under 'capacity 21'. The next meeting of senior officials shall be convened at Beijing, China, December 1993 with the support of the ESCAP, UNDP and UNEP.

The establishment of a systematic consultation network has been considered indispensable for the environmental cooperation in this region. In this sense, the Senior Officials Meeting at Seoul provided a moment because it was the very first official inter-governmental framework concerning the environment of this region. The meeting succeeded in producing agreements in the three priority areas for future cooperation in spite of the diversity of this region. It also provided an opportunity for building the foundation of sustainable regional cooperation by securing administrative and financial support from international organizations such as UNDP, UNEP and ESCAP.

4 Future Regional Cooperation and Proposals

4.1 Need to develop the Senior Officials Meeting

Given the unprecedented rapid rate of economic growth of the North-East Asian region, its environmental efforts fail to keep pace with its economic achievement. Recently, however, the environmental cooperation of this region has been activated and is heading toward official inter-governmental meetings on a regular basis, although a regular institutional mechanism has yet to be established.

One of the main considerations for regional environmental cooperation would be the mutuality of interests. Given the conflict of interests among the countries, which originate from the diversity of socio-economic systems and the levels of development, positive efforts of the relevant countries at government level will be indispensable for ultimately successful regional cooperation. Various consultations and projects at the practical level should be supported by a strong political will from each individual government. As the Senior Officials Meeting at Seoul would be the first attempt at inter-governmental level to promote regional cooperation, it would be desirable for the meeting to be charged with further consultation till an appropriate time comes for its institutional arrangement. The meeting will be able to integrate and coordinate various efforts of environmental preservation which are in force informally or sub-regionally in this region, and thereby to establish an official channel of consultation between countries.

Considering the diversity of concern of the countries, we should not be overly ambitious in its early stage. Under the circumstances where new projects have not been specified and financial and institutional arrangements have not been prepared, it will be necessary to concentrate on capacity building efforts. Thus, the exchange of environment-related information which is based on standardized method of estimation, the transfer of technology and the identification of imminent joint projects, such as joint studies, monitoring and collaborative research, will be the main issues in the discussion of the next Senior Officials Meeting at Beijing.

4.2 Institutional arrangement

So far the discussion of establishing an institutional arrangement in this region has not been activated. Given that no joint project has been implemented except for a few sub-regional ones, an institutional arrangement is not urgently required. Now, the three priority areas being adopted in the meeting at Seoul, and some projects for implementation being expected to be approved in the next meeting, a consideration should be given to creating an institutional arrangement to ensure the continuity and success of the NEAREP. To implement such a program, it is important to have some form of an institutional framework to ensure efficiency and continuity for the cooperative activities. Some projects and activities may require review from time to time, and their relative priority may change to reflect the changing regional situation. This function needs to be executed on a regular basis. An institutional body charged with this function should be established. As the NEAREP comprises more projects running in parallel, their implementation needs to be coordinated. Thus, for a mid-term or long-term perspective, the secretariat as a regional coordinating unit should be established to ensure the integrated and managed execution of various projects in this region. This concern will have to be discussed further at the next Senior Officials Meeting.

Different situations and different concerns of the countries in this region being considered, an evolutionary approach is more desirable in adopting the institutional arrangement. Until such time as an institutional arrangement is established on a permanent basis and functioning effectively, international organizations such as ESCAP, UNDP and UNEP can serve as a temporary secretariat.

4.3 Financial arrangement

It will be necessary for the countries in this region to contribute funds in order to ensure that various projects and programs are properly carried out, and in order to finance the institutional arrangement once it is established. The formula for sharing the burden by each participating country could be agreed on the basis of GNP or other factors. In general, the formula is not necessarily determined by economic principles. The internalization of externalities, which is in conformity with economic principles and is generally acceptable at national level, is hardly applicable to the case of trans-boundary environmental pollution. This issue will be considered and discussed further in the future.

Financial problems, however, should not be overly emphasized at the present stage. Given the conflict between the economic interests of the countries involved, it may make the regional cooperation slower and harder. In the early phases of regional cooperation, usually some external assistance is required to encourage a higher participation. Financial support from ESCAP for the Senior Officials Meeting at Seoul has already been made and the support from the UNDP, UNEP and the ADB is expected to be available. However, it should be recognized that the countries in this region are made fully aware of the need for financial commitments on their parts and the gradual reduction of external financial support in the long-term.

The NEAREP needs to have its own framework peculiar to the characteristics of the region for environmental cooperation. Given the diversity of socio-economic conditions, a cautious and gradual approach is recommended. For this reason, various cooperative efforts should be strongly supported by the political will of each government. The Senior Officials Meeting was a major step forward in this sense as it was the first inter-governmental meeting at the official level in this region. The future regional cooperation should concentrate on capacity building efforts in the short-term, and proceed to the establishment of an institutional arrangement and financial arrangement in the longer term.

IV.4 Lake Biwa and Sustainable Development Issues

M. Nakamura

Head of Research Division, Lake Biwa Research Institute, Shiga, Japan

Abstract

Lakes are extremely vulnerable to external disturbances, so much so that they often serve as excellent indicators of a deteriorating ecological system upland from the water body in the catchment. Throughout the history of human interaction with lakes in the Lake Biwa region, where some thirteen million people use the lake for sustaining their lives and economic activities, the issues of sustainability have arisen in various forms and have been dealt with in various manners. The management of such a great lake is often socially and politically controversial. It involves technical and economic approaches, to deal with both quantity and quality. There are many types of conflicts. The scope and the nature of issues pertaining to the sustainable development of Lake Biwa, therefore, are unique in some senses but are also representative of what is taking place and what can be expected to take place in other regions of the world where competing uses of quantity and quality resources of fresh water will affect the very means of survival.

1 Introduction

Today, in many parts of the world, the ecological and environmental integrity of lakes is rapidly deteriorating. In these lakes, much of the water resources and eco-system values have been lost.

The forms of lake environment deterioration are many. But some of the more important ones include:

(a) Excessive withdrawal of water from inflowing rivers, resulting in subsidence of lake water-level to the extent of causing serious water quality degradation and nearly irreversible eco-system alteration.

(b) Excessive development of upland watersheds through deforestation and exploitative cultivation, causing soil erosion and lake siltation.

(c) Chemical contamination of lake water and sediments caused by industrial and agricultural discharges, sometimes resulting in lasting disruption of ecological integrity by the cumulation of toxics through the food chain.
(d) Artificially induced eutrophication, often caused unintentionally through the improvement works of inflowing and outflowing channels and of coastal fringe structures, and by the construction of in-lake structures.
(e) Acidification by acid precipitations.

These and other factors, individually or collectively, contribute also to the destruction of habitats which usually result in loss of fisheries as well as the loss of indigenous species and species diversity (Figure 1).

The situations described in (a) and (b) typically take place in developing countries. The increasing population pressure leads to desperate and short-sighted agricultural and aquacultural practices. Hurriedly pursued industrialization also results in demands for clean water from and discharge of contaminated water into the nearby lakes. The lakes facilitation hydropower energy production are often subjected to induced fluctuation of water levels.

The situations described in (c), (d) and (e) are problems facing many developed countries, though some of the developing countries have also been expressing their concern. A great deal of study has been conducted in the past decades on these issues particularly in developed countries, and yet there is only a limited number of relatively successful cases of restoration.

By their very form of existence, lakes are particularly vulnerable to external forces of disturbance, so much so that they could be regarded as excellent indicators of the fragility of the mother earth. At the same time, lake management is generally a very complicated undertaking. The problem types are generally compounded and the causes and effects interrelated. Solving one problem may result in the creation of another. The general knowledge on technology, policy and institution on lake management has been impressively accumulated in the past decades particularly in Europe and the United States, but individual lake problems are unique and the solutions differ widely from one another, depending on their size, geophysical characteristics and surrounding environmental conditions, including the state of development of the watershed etc.

2 Eutrophication and its Control, a Universal Problem

Eutrophication of lakes and other enclosed bodies of water has long been a universal problem. It can happen naturally, but it usually occurs when nutrients such as nitrogen and phosphorus are artificially introduced from various land based human activities. It can be controlled, but controlling usually implies constraining human activities on land or the huge cost of removing nutrients.

In Europe and in the United States, the controlling of eutrophication has proceeded to such an extent that many lakes have now been successfully restored.[1] In other parts of the world, including Japan, the controlling of eutrophication is still a major environmental issue. The construction of a waste water treatment system for water quality protection is often unfeasible because of its cost. Even in

Japan, where municipal infrastructure development for environmental enhancement has been progressing over the past decades with huge public investment schemes, the improvement in the water quality of lakes, including that of Lake Biwa mentioned below has been only marginal. For most developing countries, it is difficult to imagine that lake water restoration will be a high enough priority to mobilize the necessary financial resources.

3 Lake Biwa in Brief

3.1 Geography and history

Lake Biwa, the largest lake in Japan, is located in the uppermost reach of the Yodo river basin in central Honshu island and occupies one sixth of the prefectural territory of Shiga. In the downstream reach of the basin are such metropolises as Kyoto, Osaka and Kobe and other smaller municipalities in Kyoto, Osaka and Hyogo prefectures, with a total population of 13 million, for which the lake supplies water for domestic, industrial, hydro-electric and irrigation purposes (Figure 2).

The lake measures 63.5 km north to south and is divided into two sub-basins at the constriction point approximately 16 km from the southern end. It has a total surface area of 674 km^3, a volume of 27.5 billion m^3 and a shoreline length of 235 km. Its catchment area is 4.7 times the lake itself and its boundary coincides pretty much with that of the lake catchment area, constituting 96% of the prefectural land.[2] The lake serves as a source of domestic and industrial water for both within the lake catchment area and the downstream population and industrial centers in the Keihanshin Area.[3] It also serves for irrigation of paddy fields in the lake basin flat lands and for hydro-power generation at some distance downstream of the outflowing river. Lake Biwa has been the most important natural asset for the prefecture, and Shiga residents have had special attachment to the lake throughout history because of their dependency on the lake for fishing and transportation.

The history of Lake Biwa and Yodo river water management has been that of conflicts of interests and their resolution between the Keihanshin Area downstream, particularly the greater Osaka region, and the Shiga prefecture upstream. For centuries the communities in the immediate surroundings of Lake Biwa had experienced many severe floodings of their agricultural fields before the government finally agreed approximately a century ago to a major dredging work of Seta River at the outlet of the lake (Figure 3).[4]

The demand for water, particularly for industrial uses, began to increase sharply as the country entered the era of economic growth a decade or so after the end of the Second World War.[5] When the Japanese economic growth gained momentum in the mid-1950s, there were surging interest from the downstream population and industrial centers in the development of a comprehensive development plan of Lake Biwa water resources. After nearly two decades of political pressure put on the Shiga prefecture, consisting of demand by the downstream interests and the initiatives by the national government ministries, the Shiga prefectural government finally agreed to a scheme for the development of Lake Biwa water resources, called the Lake Biwa Comprehensive Development Project (LBCDP), involving massive construction works around the lake.[6]

In the meantime, the lake water quality has been worsening over the past several decades due to population growth and various development activities within the catchment area. Some environmentalist groups have been putting the blame on LBCDP itself. The Shiga prefecture is now under pressure to improve the lake water quality through various wasteload reduction measures as well as in-lake clean-up attempts.

3.2 Lake characteristics and management issues

Few lakes in the world possess the intricate history of human-nature interaction which characterizes that of Lake Biwa, the largest and, in virtually all possible aspects, the most important lake in Japan. To understand the lake, its limnology, ecology and environment, socio-economic implications, resource values and all, one would almost have to live at the lake and sense the weight of its history. Lake Biwa today reflects centuries of humans' unique affairs with the lake, and it also reflects, naturally, what it will bequeath to his future generations.

The most important facts abut Lake Biwa could be summarized as follows:

(1) The largest and the most important lake in Japan. It is the largest lake both in area and in volume and it has been, throughout modern history, the single most important water source in the western part of Japan.

(2) The Shiga residents' attachment to the lake. For its irreplaceable tangible as well as intangible natural resource worths, the prefectural residents have historically shown extreme attachment to an extent perhaps unparalleled by any other lake cases in Japan.

(3) Relatively short history of lake management efforts. It has only been a century or so since construction of major control facilities for water resource development and flood control were first undertaken, and it has only been a little more than a few decades since the prefectural government began to engage in serious and systematic pollution control activities.

(4) The intricate policy decisions. The history of the development and conservation of the lake catchment area reflects extremely intricate policy decisions, sometimes resulting from severe conflicts of interest among national, regional and local entitites at various levels of decision hierarchy.

(5) Complex interactions among the concerned bodies. There emerge, therefore, many complex structures of interaction among major role-players, such as governments, industries, citizens and others, both with respect to the development and implementation of development and conservations policies involving the lake and the catchment.

The important subject areas requiring analysis for a thorough understanding of the evolving issues in Lake Biwa water quality management include:

(a) The national, regional and local legislations which provide the basis for economic development, land use and land development, water quantity/quality management.

(b) The legal framework for land use control and its relationship with the development of major water pollution control programs.
(c) The water use trends and the development of locally appropriate water management systems, particularly for agricultural water use and run-off control.
(d) Major water quality conservation measures for the control of domestic, industrial and agricultural pollution loads.
(e) The making of the eutrophication control ordinance, particularly with respect to its initial driving movements, its legal, social and policy implications and the impacts it has generated both in terms of actual control of eutrophication and of socio-political significance.
(f) The intricate mechanisms of interaction between the local government agencies and the prefectural residents in dealing with lake water pollution incidents.
(g) The Lake Biwa Comprehensive Development Project (LBCDP), its development process, contents and progress, and its implications and impacts.

4 Facilitating the Development Pressure: Altered Land and Water Uses

Prior to making observations on the development and conservation dynamics which will be discussed in the following sections, it would be appropriate to review the relationship between the economic development impetus and the resulting changes in water quality as well as the process fulfilling water quality and quantity requirements for development activities. The discussion will center around two basic issues: land and water uses.

4.1 Land use

In the Lake Biwa catchment area, a major part of the flat lands surrounding the lake have historically been dedicated to rice growing lands. The land use patterns began to undergo noticeable change first in the 1940s. The filling of lagoons around Lake Biwa took place right before the Second World War and immediately afterwards, primarily for the purpose of increasing rice production, though this never really happened. The filling of parts of the South Basin of the lake itself took place in the 1960s, basically for converting wet lands into arable and/or inhabitable lands (Figure 4).[7]

Many projects to convert farm land into industrial estates and residential establishments took place in a rather piecemeal fashion during the periods of intensive industrialization and urbanization from the early 1960s to the mid-1970s. The combined effect of these changes was the uniquely Japanese sprawling of the land which had once been dedicated to paddy-field cultivation.

The forest area covers a little more than half of the entire prefectural land today. In the medieval age, a typical problem associated with forest management in the Lake Biwa watershed was deforestation.[8] In the last few decades, however, the problems associated with forest land have become more subtle. As in the case of agricultural lands (paddy-field and other crop lands), the forest fringe lands have slowly but steadily been cleared and graded for various developmental purposes including housing and industrial estate development. In the process, the forest land has slowly lost its capability to function as a vast reservoir of precipitated rain and snow waters, and nutrients contained in the precipitation are run off into nearby streams. It is said that the amount of nutrient discharged from forest land today amounts to almost as much as that discharged from agricultural fields.

Upon introduction of the National Land Use Planning Law,[9] the prefectural government was given the opportunity to correct the situation and realign fragmented land parcels dedicated to many different uses. But the lands, once allocated for different uses, have been difficult to convert back to uses consistent with the desired planning framework. It may be said that developing of land use schemes that are consistent with the existing land uses and appropriate to future development goals of the prefectural government will continue to be a challenge for the prefecture.

4.2 Water use

One of the most outstanding features of Lake Biwa water use and allocation today is that of the water withdrawn from the lake. The agricultural sector has the biggest share, amounting to 93% of the total, despite the fact that the prefecture has undergone industrialization and urbanization on a rather significant scale in the last few decades. Throughout history, the agricultural sector has had the oldest, and most complex water use practices and well established customary rules of withdrawal. The industrial and municipal use demands which began to increase in the 1960s had to compete against agricultural demand. The resulting impacts were twofold; the establishment of a new rule to supersede the current destabilized and ad-hoc water use and allocation rules, and the need for serious effort for the control of point as well as non-point pollution. It also became clear that there is a need to balance water uses where they compete for scarce water.

(a) Agricultural water use

Agricultural waters have been obtained through history in one of three ways. At or near the foothills, particularly on the eastern side of the lake, there are a number of irrigation ponds.[10] Irrigation in the flatlands, however, depends on either or both groundwater pumping and lake water withdrawal, the latter being particularly prevalent in the delta regions of major river systems.[11]

As agricultural water demand increased after the Second World War, it became necessary to ease severe water allocation conflicts between upstream and downstream communities in the lake catchment area. Many dams of significant scale were constructed along the river systems. At the same time, the use of lake water for irrigation became possible even in regions distant from the lake, as extensive irrigation network systems were constructed. The network in effect created water flows in the reverse direction. As a result, the areas previously devoted to groundwater irrigation began to shrink. The combined effect of the expansion of the lake water irrigation area in the upstream direction and of river water irrigation area in the downstream direction was serious lake water pollution, phasing out the previous repetitive water use through dendritic or cascade channels.

(b) Industrial water use

Industrial water use has been dependent on industrial types and sizes. The early days of industrialization in Shiga were very much dictated by textile industries, particularly of a synthetic silk thread called rayon. Today, major industrial categories in the prefecture are inland assembly industries, which require water in only limited quantity. The amount of water used by all categories of industries is 1.36 million m^2 per day, of which textiles, chemicals (pharmaceuticals), plastics, potteries, electro-machineries and transportation machineries are the major water consumers.[12]

They are also the most likely sources of point source pollution, and the effective control of waste-load discharge from these industries will become the key to achieving significant reduction in water pollution. There is a very important link between agricultural and industrial water uses, in that the implementation of large-scale irrigation systems has greatly increased the opportunity for groundwater contamination, resulting in serious damage to the operation of textile industries requiring uncontaminated groundwater. On the other hand, the groundwater users have been avoiding direct confrontation with surface water users who pollute groundwater by continuing to extend the well depth to reach deeper and cleaner aquifers.

(c) Domestic water use

Drinking water supply has its own history. Prior to the introduction of piped water supplies, a major portion of drinking water was extracted from nearby streams and wells. Today, approximately 60% of water supplied by domestic water supply systems originates from Lake Biwa.[13] The single most important factor affecting the increase in domestic water supply has been the population migration throughout the last several decades from the Keihanshin (Kyoto, Osaka and Kobe areas) region into the prefectural jurisdiction. The provision of water to these new migrants was a precondition to facilitating housing development activities, which had to be an unavoidable consequence of the industrialization and urbanization of the prefecture.

It is said also that the traditional water usage in the lake front communities were quite efficient in that there were self-regulatory efforts made to prevent the used water from being directly discharged into streams flowing into the lake. The used waters were held for a period in a small holding pit prior to discharging into the receiving stream. In addition, the amount of used water generated by households was much less when there were no water supply systems, as most washing was done directly at water fronts. When piped water supply systems were introduced into the agricultural and/or fishery communities, they effectively destroyed the traditional style of water management. Now that a large percentage of the population depends on lake water through piped water supply systems, lake conservation and lake water use have become synonymous to many Shiga residents. This has brought an interesting situation with respect to the Shiga residents and the residents of downstream prefectures, in that they now share a common concern and sentiment about the future of the lake water quality, which has never been the case before. At the same time, those engaged in agricultural and industrial sectors have been placed in a situation in which they will have to face the consequences of their own deeds in having foul water quality.

5 Coping with Environmental Stress: an Assortment of Conversation Efforts

As stated above, the Lake Biwa catchment area saw significant changes take place as a result of insurmountable development pressure exerted by the national development policy, leading to intensified development activities downstream of the lake system. The most obvious outcome was a sudden surge of environmental stress exerted on the lake, particularly on its southern basin (the South Lake). For centuries Lake Biwa has had no large-scale i-lake aquaculture farms, such as fish pens and fish cages; it is fair to say that most of the nutrients accumulated in the lake have originated on land and have been transported through rivers and streams. For centuries, or for millions of years for that matter, the lake as a whole maintained a fairly stable ecological balance, taking advantage of its vast self-purification capacity. Being a confined body of water, despite its huge water volume, the deterioration of water quality and hence its disturbed ecological balance was suddenly to become a reality, once the pollution load inputs exceeded a threshold. In terms of geological time, it was almost an instantaneous event for Lake Biwa. The balance sheet of pollution, therefore, was of prime interest to those concerned. Who is (was) responsible? What actually happened?

5.1 Understanding the state of the lake

Most amazingly, even today, this question has not yet been answered in scientifically and convincingly precise terms. It is known that, in terms of organic pollution loads generated on land, the deteriorating water quality could be attributed approximately equally to domestic, agricultural and natural sources. However, scientific explanations of exactly what triggers physical, biological and chemical interaction of the natural and artificial substances which result in the eutrophic state of the lake, are at most sketchy.

Further, the pollutant generation dynamics on land have been undergoing tremendous changes even over the pat decades, making it even more difficult to be scientifically correct. There is a greater number of people and industries resulting in the generation of more pollutants. Industrial point sources such as discharges of toxic substances have been under control, thanks to the introduction of stringent regulations and rather severe enforcement efforts since the 1970s. But the discharges from small to marginal industries and from households have been on the increase.

Today, the discharge of gray-waters from individual households as a result of the absence of sewerage systems is said to be one of the major contributors to lake pollution. There have been changes in agricultural practices also. A great majority of paddy-field farmers today are weekend farmers, resorting to highly mechanized production practices and generous use of fertilizers and pesticides, which was not the case 30 years ago. Thirty years ago, the individual paddy-field units had semi-naturalized paddy-field channels connecting individual units in which a significant degree of trapping of outflowing nutrients took place. But after the institution of the paddy-field land reform project, highly streamlined inflow and outflow control channels were constructed, and they are said to be in part responsible for a larger quantity of nutrient run-off than before.

In all, an assortment of conservation measures had to be introduced to cope with the rapidly deteriorating environment which had been subjected to extreme development pressure exerted on a large portion of the catchment area. a brief outline of these pollution control measures is presented below for domestic, industrial and agricultural waste water, together with a discussion on water quality monitoring and the Eutrophication Control Ordinance of 1979.

5.2 Coping with waste-load reduction

Of the three categories of waste waters, management of the domestic waste water has attracted the greatest publicity in the recent past, as the Shiga residents realized that the planned construction of sewerage systems around the lake would take quite some time, and that by then the lake may have already been irreversibly eutrophied. According to the latest statistics, only some 30% of the population has been provided with service through several public sewerage systems.

There are four regional waste water collection and treatment systems, of which two are partially completed and in use. They will eventually take in a major portion of domestic waste water and part of non-toxic industrial waste water. The remaining percentage of the population must be served by one of several waste water management systems, including a night-soil collection and treatment system and on-site household treatment systems of various kinds.

Thus household owners in those areas resort to an on-site treatment unit. Most household systems marketed in the past allow only toilet waste water to be treated. The gray-water, a term used to indicate kitchen and bathroom waters collectively, makes its way out into ditches then to streams and eventually to the lake through river systems.

Industrial waste waters are generally regarded to have been brought under tight control, particularly after the institution of the Water Pollution Control Law. There has not been serious violation of the law as in the past, thanks to very stringent enforcement, the awareness of the public media and the citizens to any sort of spill incidents, and the realization on the part of the industries that because of high penalties illegal discharge of pollutants will not pay off. The discharge of substances potentially harmful to the environment and to public health is regulated by the Water Pollution Control Law. In addition, discharge of nutrients which cause lake eutrophication is controlled by the Eutrophication Control Ordinance of 1979 and the Clean Lake Act of 1984.[14]

Agricultural waste water, including dairy waste water, is also regarded as a serious lake polluter, although the approaches for pollution control differ from such point sources as domestic and industrial dischargers. As for paddy-field cultivation, pesticides and fertilizer uses are of greatest concern. The changing agricultural practices for improving productivity in the recent past have yet to provide the solution for improving the efficiency of pollution control. Attempts are being made to improve the application of pesticides and fertilizers, both in terms of management practice and technological innovations. Point injection of fertilizer at the root of the rice seedlings is being tested and known to be gradually accepted by farmers. Recycling irrigation water, drawn up into the irrigation network by large pumps, is also being proposed and is under investigation for its technical and economic feasibility. In addition, the environmental monitoring center conducts paddy-field discharge monitoring at spring cultivation and fertilizer application season to check incidents of gross pollution.

5.3 Significance of hierarchical legislative structure

Japan is a nation with a highly hierarchical decision making structure, and the central government sets the basic policy direction in most of the issues of general national concern. Setting the national and regional development policy, of course, falls right into this category of hierarchical decision making. Development plans for regions and prefectures will have to be prepared closely in response to the framework established within the long-term national development plans.

Shiga's own comprehensive development plans of 1963 and 1967 had great impacts on the profile of the Lake Biwa catchment area, which resulted in part in the gradual deterioration of lake water quality.

Water management legislation may be categorized into legislation on the management of rivers, forests and seashores, including landslide and flood control, on water resources and water use and on water quality conservation. The categories of water legislation, just as in the case of the legislation of regional development, exhibit a very strong flavor of the centralization of management authority. The national government, for example, has taken initiatives to provide the broad framework on pollution control by enacting the necessary legislations on the protection of water, air and other environmental resources.[15]

It is to be noted, however, that there has been a significant number of cases dealing with environmental issues where local governments took an initiative in establishing the basic framework of legislative action. The Eutrophication Control Ordinance of Shiga prefecture mentioned earlier is one of the most outstanding examples of this.[16]

5.4 Eutrophication control ordinance

These pollution control efforts are not unique to Shiga prefecture. If anything is unique about pollution control in the Lake Biwa catchment area, it must be the spirit and symbolism reflected in the Eutrophication Control Ordinance enacted after the incidence of the first red tide in 1977. Perhaps the most important aspect of the whole sequence of events leading to the enactment of the ordinance and its implementation afterwards, were the citizen initiatives to refrain from using synthetic detergents containing phosphorus builders (discussed in Section 7.2).

At the height of the movement immediately after the enactment of the ordinance, almost 90% of the households surveyed were willing to collaborate and not use detergents containing phosphorus. With an assortment of provisions in the ordinance, including regulations on industrial discharges of phosphorus and nitrogen containing waste waters, the water quality of the lake seemed to show slight improvement within the first few years. However, the lake water quality has since remained more or less the same. Various land development activities and continuing inflow of non-point sources of pollution are blamed as the major contributing factors.

6 Development and Conservation Dynamics

6.1 Dynamics of development efforts

The dynamics of development efforts in the past several decades could be described in many different ways. However, one of the most obvious and most important forcing functions has been, of course, the major policy decisions of the national government. Throughout its modern history, the Shiga prefectural government has, as have the other prefectural governments, been subjected to strategic regional development programs under the national umbrella development policy framework. These programs have collectively destined the nation to its economic achievements in the post-war decades.[17] In effect, the dynamics of development efforts have reflected the trend of the time, implying that local development efforts had to be responsive to national and regional development needs as well as local inclinations.

Of course LBCDP constitutes an important part of the development dynamics. It is closely linked with the framework of regional development mentioned earlier, but it could well be regarded, historically and nationally, as a very unique undertaking for consolidating the social overhead capital specially directed to an unsurpassable natural resource of the nation, Lake Biwa. How it fares in the formula of the dynamics of development efforts is not simple to analyze because of the difficulties in assessing its socio-political implications and in quantifying economic and social benefits.

Another important observation on the development dynamics is that, naturally, development efforts result in impacts of all kinds on all phases of development activities, including industrial structuring, social and demographic alterations, land and water use changes and increasing environmental stresses. The industrial structure has changed over the past four or five decades, allowing the secondary and tertiary industries to overcome the primary industry in terms of industrial outputs. The social and demographic alterations, such as changes in life-style, community structure, labor force structure and even cultural and educational opportunities and facilities, are of great significance in characterizing the societal dynamics. The land and water use changes have already been elaborated in the previous section. But the fact that changes took place, in an extremely short span of time in terms of the development history of the catchment area, should be taken most seriously.

6.2 Dynamics of conservation efforts

The dynamics of conservation efforts in the Lake Biwa case are unique on the one hand and typical on the other. They are unique because Lake Biwa has had unique problems with respect to its geography, hydrology and socio-economy. The conservation measures therefore had to be responsive to these unique elements. It could also be called typical in that many of the conservation measures introduced have been typical of the particular time in Japan and of the manner in which the government and the citizens have dealt with the impending issues.

The causal relationships among Lake Biwa undergoing eutrophication, the public reacting to deteriorating water quality and the local government attempting to respond to such reactions may be illustrated using the diagram shown in Figure 5. Some of the factors associated with each of the causal relationships are described below for two succeeding decades, one for 1970-1980 and the other for 1980 to date. During the former period, the rapid deterioration of lake water quality prompted highly-charged citizen concern and interest, which in turn became instrumental for the introduction by the government of various remedial measures to achieve eventual restoration of the Lake Biwa water quality from the state of eutrophication. During the latter period, it became more apparent that the lake, once having become eutrophic, would be extremely difficult to restore in a short period of time, and that the introduction of various remedial measures for reducing waste-loads or for recovering polluting agents from the lake can only facilitate, not resolve, eventual recovery of lake water quality.

1973–1980 Period

(1) Causes of environmental disturbance:
- intensification of industrial activities;
- increase in population migration from downstream regions;
- development activities including those specified in LBCDP.

(2) Apparent deterioration of environmental integrity:
- disturbance of plankton calendar resulting in large-scale red tides;
- large-scale lake shore filling and the resulting reduction in natural lake fronts.

(3) Impacts and damages to public facilities:
- clogging of water treatment filter-beds;
- closing of bathing beaches.

(4) Enhanced citizens' awareness on environment:
- sense of alarm about the deteriorating quality of drinking water;
- grievances for the loss of opportunities for water front activities.

(5) Message from environment to citizens:
- taste and odor problems;
- increased turbidity and algal growth.

(6) Citizens' demand on government:
- regulation of discharges of nutrients from industries;
- moratorium on development activities;
- provision of inexpensive sewerage service.

(7) Citizens' initiatives:
- use soap movement;
- efforts to reduce household pollution loads.

(8) Message from environment to government:
- high N and P concentration in lake waters;
- occasional algal blooms.

(9) Government action:
- enactment of eutrophication control ordinance for regulating the use of phosphorus containing detergents;
- monitoring of industrial waste water discharges;
- control of agricultural run off control;
- sewerage construction.

1980 to Date

(1) Causes of environmental disturbance:
- non-point sources of pollution, specifically from agricultural sources and from land development activities for urbanization;
- non-specific sources of pollution, including discharges of household gray-waters.

(2) Apparent deterioration of ecological integrity:
- continuing incidents of algal blooms, including those by Uroglena Americana (red tide), Microcystis (greenish growth) and pico-plankton (plankton of a size close to bacteria);
- declining catch of indigenous fish.

(3) Impacts and damages to public facilities:
- installation of activated carbon filter beds for removing plankton induced trace organics causing taste and odor problems;
- introduction of beach cleaning machineries for removing floating weed debris washed ashore.

(4) Enhancement of citizens' awareness on environment:
- strong sense of loss of 'good old days' of clean streams and drinking water free of taste and odor;
- growing concern about the discouraging sign of gradual deterioration rather than improvement of lake water quality;
- linking of Lake Biwa issues with those of the global environmental deterioration.

(5) Message from environment to citizens:
- no sign of water quality improvement despite various efforts;
- continuing taste and odor problems.

(6) Citizens' demand on government:
- curtailing of coastal development causing increased run-offs of land based pollution;
- introduction of non-point source pollution control measures;
- litigation against LBCDP stating that it is the culprit of eco-system destruction.

(7) Citizens' initiatives:
- various voluntary initiatives for clean-ups and life-style changes;
- establishment of an 'environmental cooperative';
- establishment of an 'environmental accounting consulting firm'.

(8) Message from environment to government:
- N and P concentration at stand still;
- first sighting of pico-plankton.

(9) Government action:
- development of comprehensive environmental management plans;
- intensification of environmental education;
- promotion of basic limnological research;
- enactment of reed-bed protection ordinance.

Many interpretive observations could be made about this simple diagrammatic presentation. For this paper, however, let us confine ourselves to observations involving citizens and their initiatives.

First, the citizen initiatives for environmental conservation took many different forms in the Lake Biwa region. The initiatives have been taken by different citizen groups, at different geographical locations, for different reasons. Some have achieved their aims and objectives, while others have failed to do so. Some have had lasting impacts on shaping social value in the region, while others didn't. Some have been more spontaneous than others. Some could be considered unique to the region, because the issues dealt with were strictly local and pertained exclusively to the Lake Biwa situation. And yet many could be considered otherwise, because the problems pertaining to such initiatives were rather universal to the nation.

For example, do the residents in the Lake Biwa catchment area feel a greater sense of environmental responsibility and take firmer initiatives than those in other areas in Japan? Perhaps so, because living in the area automatically carries with it a sort of social pressure to care for the lake. The schemes evolved through citizen initiatives, such as the use soap movement (instead of phosphorus containing detergents), have been praised as epoch-making. The gradual erosion of interest in the movement over a period, however, seems just as typical as any other citizen-initiated actions elsewhere. It is difficult therefore to characterize the citizen initiatives in the Lake Biwa regions as something unique and worth of special treatment.

Secondly, it is true, however, that conservation initiatives have evolved rather prominently in the region, with a unique mix of different motivations, interests and capabilities on the part of the citizens, the government and the environmental professionals. There may be many reasons for this.

One of the reasons lies in the nature of the lake itself. In the case of Lake Biwa, it is too important for each and every interested body not to have a say or action about its presence. The greater the stake, the tenser the relationship gets. And, as with the basic principle of democracy, such a tension is basically healthy. Another reason may be that there has been a unique blending of sentiment between the urban dwellers who immigrated to the lake watershed and have a recent attachment to the great lake, and the long-time Shiga residents for whom the lake has for centuries been a part of their life. The recently immigrated urban dwellers are often more vocal and active in taking environmental initiatives, having nothing to lose in claiming clean water and amicable water fronts. On the other hand the long-time dwellers in old communities are more careful, because they live and have lived at the expense of exploiting the land, water and other natural resources in the Lake Biwa region.

Regardless, the fact that the lake catchment area and the prefectural jurisdictional area almost totally coincide is a unique contributing factor to the prefecture fostering such a sentiment. The people living within the catchment area are constantly being reminded that they have their own role to play in endeavours to achieve environmental integrity. And this has precipitated unintended 'partnership' between the residents and the government and among the residents themselves.

Thirdly, despite the above, the approaching trend appears to be for greater involvement of and genuine concern by the citizens. A so-called environmental cooperative was established in Shiga a few years ago, a first of its kind in Japan, drawing much attention and interest as a very significant new environmental movement. Similarly, for the first time in Japan a business venture offering services of environmental accounting to business establishments was recently set up in Shiga. There are other firsts, including those pertaining to environmental education. Thus, there seem to exist some sort of synergistic forces working behind all such initiatives.

It may have a lot to do with the fact that Shiga is becoming a melting pot of environmental scientists and environmentalists. With the advance of information technology, a group of capable individuals living in the Lake Biwa region could easily take advantage of the symbolic importance of the lake to become a center of information generation and dissemination on the environment. Working with a large number of citizens, including school children, requires not only symbolism but technology and scientific knowledge. Undertaking mass environmental education requires a critical mass of interested and capable individuals.

Finally, while the contributing factors to the dynamics of conservation efforts have changed over time, the basic structure and the substance of the dynamics has remained fairly intact. Though new attempts and new developments have been introduced to the lake restoration efforts, the lake water quality has not shown any encouraging sign of improvement. Though the relationships between interest groups, communities and government agencies have evolved to be more mature, there still remain conflicts and unresolvable basic differences in the proposed management approaches to Lake Biwa. But all in all, it comes down to the lake versus humans, or the basic issue of sustainable development.

7 Concluding Remarks

The Lake Biwa development and conservation policies have been formulated in response to various national and regional initiatives and requirements. Thus, a wide array of policy organs have in some manner or shape participated in their formulation. These organs could be categorized according to their governmental hierarchy, such as national versus local governments; governmental sectors of responsibility, such as agriculture versus industry; regional economy and intra-region political positions, such as Shiga versus downstream prefectures; economic zones, such as Keihanshin versus in-prefecture areas of concentrated industrial establishments; or even according to the ascribed basic attributes of political influence, such as the bureaucracy versus the citizens. Each such organ has played its role in pursuit of its specific objectives.

It is to be noted, however, that the intricate dynamics of both development and conservation effort, and the involvement of a variety of actors, have resulted from many centuries of rich history of human-nature interaction in the area which has been so very important socially, economically and politically. While interactions of actors of the type described in the Lake Biwa case may seem typical for cases involving any important water resources, the Lake Biwa case is unique in many respects. The very long history of human-lake interaction has created indigenous institutions and management practices, aroused particular sentiments for and/or against each other and provided a unique sense of balance among the concerned actors. A complex mixture of practices, sentiments and the sense of balance has resulted in what appears to be the very unique Lake Biwa experience.

Acknowledgment

This paper was partly prepared using previously prepared texts by the author, including those prepared for The World Bank Seminar on Economic Development and Environmental Management, 15 November – 5 December 1992, Kota Kinabalu, Malaysia, and CIESIN (the Consortium for International Earth Science Information Network, NASA) Water Institute, held at Saginaw Valley State University, Michigan, USA, 12–16 July 1993. Also, the original text of some of the sections are excerpts from the Lake Biwa Case Study conducted for UNCRD/ILEC/

UNEP Expert Group Workshop on River/Lake Basin Approach to Environmentally Sound Management of Water Resources, listed in the references below. I am deeply indebted to the Case Study members with whom I had the privilege to work from 1987 to 1990.

Footnotes

1 For example, Lake Malaren of Sweden, with Stockholm located at its outlet, is served by some 160 waste water treatment plants which collectively remove 90% of inflowing nutrients. The lake now maintains excellent quality of lake water which once was eutrophic. In the United States, Lake Washington and Lake Tahoe present themselves as famous success cases. In the case of the former, it was the diversion of treated waste water through a 180 km long pipeline which lead to the recovery of water quality from the once eutrophic state. In the case of Lake Tahoe, it was a combination of waste water treatment and good management of land use and recreational activities, which has resulted in the protection of the lake in a nearly pristine state.

2 The catchment area consists of forest covered hills and mountains (60%), paddy-fields and other farm lands (25%) and urban and industrial areas. The prefectural capital, Otsu, has an approximate population of 240,000 and is located at the southern end of the lake. Some 400 rivers and streams flow into the lake, but there is only one natural water course flowing out of the lake, Seta River, and two canals passing lake water to Kyoto. The official designation of the whole of the Yodo-Uji and Lake Biwa water bodies is the Yodo river system. Its total catchment area, as referenced at Hirakata, some 20 km upstream of the river mouth, is 7,281 km^2. Its annual average flow, high flow and low are, respectively, 177.6 m^3/sec, 226.8 m^3/sec, and 117.0 m^3/sec.

3 At present metropolitan regions of Osaka, Kyoto and Kobe depend almost exclusively on Yodo river industrial and municipal water supplies. This being the case, Lake Biwa accounts for water supply amounting to 20 billion tons per year serving as many as 13 million people living in the Keihanshin area including Shiga. The lake pollution, therefore, has been a serious concern, not only for those living around the lake, but also to those receiving water from the Yodo river. In the past century or so, the water use pattern has changed significantly as a result of various water resource development activities. The two canals linking the lake and Kyoto were constructed in 1890 and 1912, respectively. The city of Kyoto, therefore, has had direct access to the lake water unlike other downstream metropolises.

4 This dredging was carried out in combination with the construction of Seta Weir, the only artificial water flow control facility of the lake outflow which is located just a few kilometers downstream from the lake outlet. The weir, constructed in 1905 and renovated in 1961, controls the lake water level and discharge rate to the Yodo river. The flood frequency and the flood damage in the lake catchment have drastically reduced since the weir construction.

5 In the downstream stretch of the Yodo river, the Hanshin industrial belt established in pre-war years began to thrive with unfulfilled thirst for more water. Exploitation of groundwater soon became constrained because of competition among industrial establishments for its use and because land subsidence caused by overdraft of water. Industries were then forced to look for alternative sources of water. Domestic

water supply needs also began to increase in the Yodo river area after suburban cities joined Osaka for gaining access to the Yodo river water.

6 In the Lake Biwa Comprehensive Development Project (LBCDP) inaugurated in 1972, the development of a new water resource amounting to 40 m^3/sec has been agreed upon by Shiga and the downstream prefectural and municipal governments, as well as by the national government. This additional amount of water was to be made available upon completion of all the lakeshore reinforcement works as it was originally scheduled and other compensatory public works projects for the Shiga prefecture by the end of 1982.

After experiencing difficulties in implementing an array of component projects within the originally scheduled 10 year project period, the project term was extended for an additional 10 years up to the end of 1992, with integration of additional projects for environmental conservation. By the end of 1992, almost all water resource development projects will have been completed, but not all of the compensatory public works, leaving the Shiga prefectural government to propose to release for the downstream communities only 27 m^3/sec of additional water on an as needed basis. The downstream governments reluctantly agreed to the tentative arrangement.

LBCDP involves such project categories as water resources development, flood control and nature and environmental conservation. An array of large-scale component construction works, costing the region and the nation billions of dollars, have been and will continue to be undertaken until the end of 1996 when the term of the project period is to terminate.

7 What appears to be gross mismanagement of land began to result at about this time, principally due to the introduction of a new set of transportation routes enabling the south-eastern part of the lake to be included in the marketing networks of the Keihanshin industrial region. The outcome of this change was haphazard exploitation of the then prevailing agricultural farm lands.

8 For example, Tanakami mountain range in the south-eastern edge of the watershed has long been extensively cleared by timber logging operations used for constructing temples and shrines in Nara and Kyoto area centuries ago. More recently, growing population pressure and the demand for firewoods has resulted in exhaustion of deciduous forests particularly during the period when Japan was nearly totally secluded from the rest of the world (1639-1854). These exploitative activities have contributed to denuding of forests and the resulting severe flooding and siltation problems along many of the Lake Biwa river systems.

9 The basic framework on the use of national land is dictated by the National Land Use Planning Law of 1974. To overcome the shortcoming of land use planning practices prior to the enactment of this law, the law stipulates the need for setting overall land use schemes, superseding, if necessary, any existing plans established within the framework of sectorial legislations.

10 The pond waters were the only source of water for centuries in the high altitude fields and the fields distant from the lake. They are still used today, but in combination with other means of water supply and basically for auxiliary purposes.

11 Lake water irrigation has evolved over centuries of agricultural practices. For a long time until electricity became available, creek water had been pumped or scooped by manual labor. In the Meiji era (1868-1912), the lowered lake water level necessitated direct water withdrawal from the lake by electric pumps.

[12] With respect to the water use pattern by industrial size, those with between 100 and 300 employees use the greatest amount of water. Many of these industries do not have the financial resources to invest in water recycling, particularly because they are not large enough to enjoy the economies of scale.

[13] Today there are six municipalities, 26 townships, one water utility agency and one association engaged in undertaking water supply, and 97.5% of the prefectural residents are supplied water through piped water supplies. Currently there are two regional water supply systems, both of which withdraw water from Lake Biwa.

[14] Aside from the established effluent standards under these legislations, there are permit systems for industrial operations. The categories of industries engaged in production activities which have the potential of inadvertent discharge of toxic materials are subjected to a permit system. Similarly, those industries or premises discharging waste water in quantity are also subjected to a permit system, together with once a year surveillance and imposed self-monitoring of discharge waste water quality.

[15] Of a series of legislations concerning environment, the ones with foremost importance are The Basic Laws of Pollution Control (1967 revised in 1970) and The Water Pollution Control Law (1970), which together marked the dawn of the era of environmental conservation after a few decades of near total disregard to environmental conservation in the name of post-war reconstruction and miraculous national growth by means of industrialization. Many local legislations on environmental conservation were enacted following these two key legislations. The Revised Shiga Prefecture Anti-Pollution Ordinance of 1972, for example, has its basis on the Revised National Basic Law of Pollution Control of 1970.

[16] The first of the post-war anti-pollution legislation was not a national legislation but it was the ordinance on the control of industrial pollution enacted by the Tokyo Metropolitan Government.

[17] The regional development efforts are facilitated through appropriate legislations, its principal legislation being the National Comprehensive Development Law of 1950. In the course of nearly 40 ears since its enactment, four national comprehensive development plans were prepared. In the Shiga prefecture, six prefectural development plans of major significance were prepared, the current plan being prepared in 1987. A whole series of development activities took place in the 40 year period, the most important ones being the completion of major transportation arteries, such as Route No. 1 Highway, New Oohashi Bridge, and Oomioohashi Bridge. In addition, industrial estates of varying sizes either moved in from the Keihanshin region or were newly established particularly along the south-eastern edge of the lake.

References

Lake Biwa Case Study Team, 'Lake Biwa Case Study of 1st Year, A Historical Review of Resource Conservation and Exploitation of Lake Biwa Environments', UNCRD/ILEC/UNEP Expert Group Workshop on River/Lake Basin Approach to Environmentally Sound Management of Water Resources, 8–19 February 1988, Otsu and Nagoya, Japan

Lake Biwa Case Study Team, 'Lake Biwa Case Study of 2nd Year, Evolution of the Lake Biwa Water Resources Management and Environmental Conservation Policies', UNCRD/ILEC/UNEP Expert Group Workshop on River/Lake Basin Approach to Environmentally Sound Management of Water Resources: Focus on Policy Responses to Water Resources Management Issues and Problems, 16–25 January 1989, Bangkok and Hat Yai, Thailand

Masahisa Nakamura and Michio Akiyama, 'Evolving Issues on Development and Conservation of Lake Biwa—Yodo River Basin', Water Science and Technology, Vol. 23, Kyoto, pp93–103

Masahisa Nakamura, 'Comprehensive Development of Lake Biwa', UNCRD/ILEC Symposium on Water Resources Management with the Views of Global and Regional Scales', November 18–20 1991, Otsu, Japan

JICA (OITC), ILEC, 'Appropriate Technology and Measures for Lake Environment Conservation', prepared as a course note for JICA/ILEC Training Course on Lake Water Quality Management, 1991.

Figure 1
Relationship among factors contributing to deterioration of lake environments

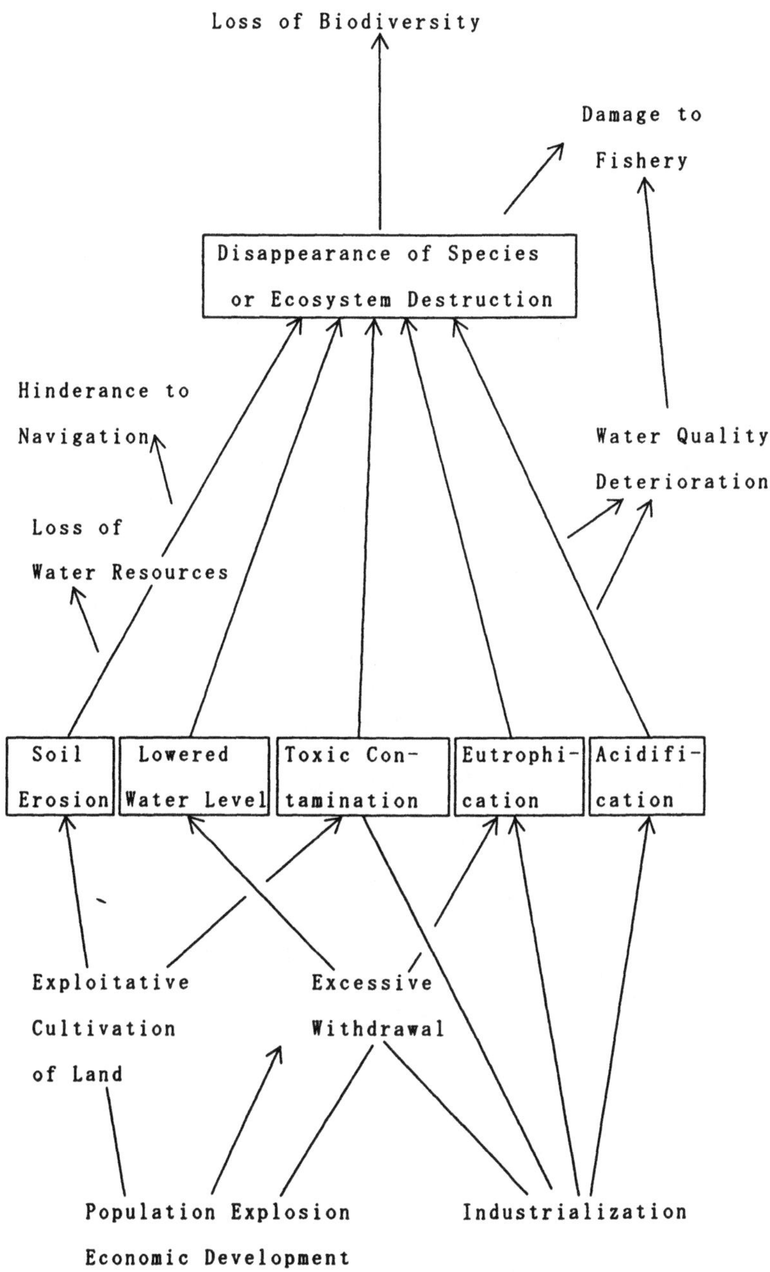

Figure 2
Lake Biwa and the Yodo River Region

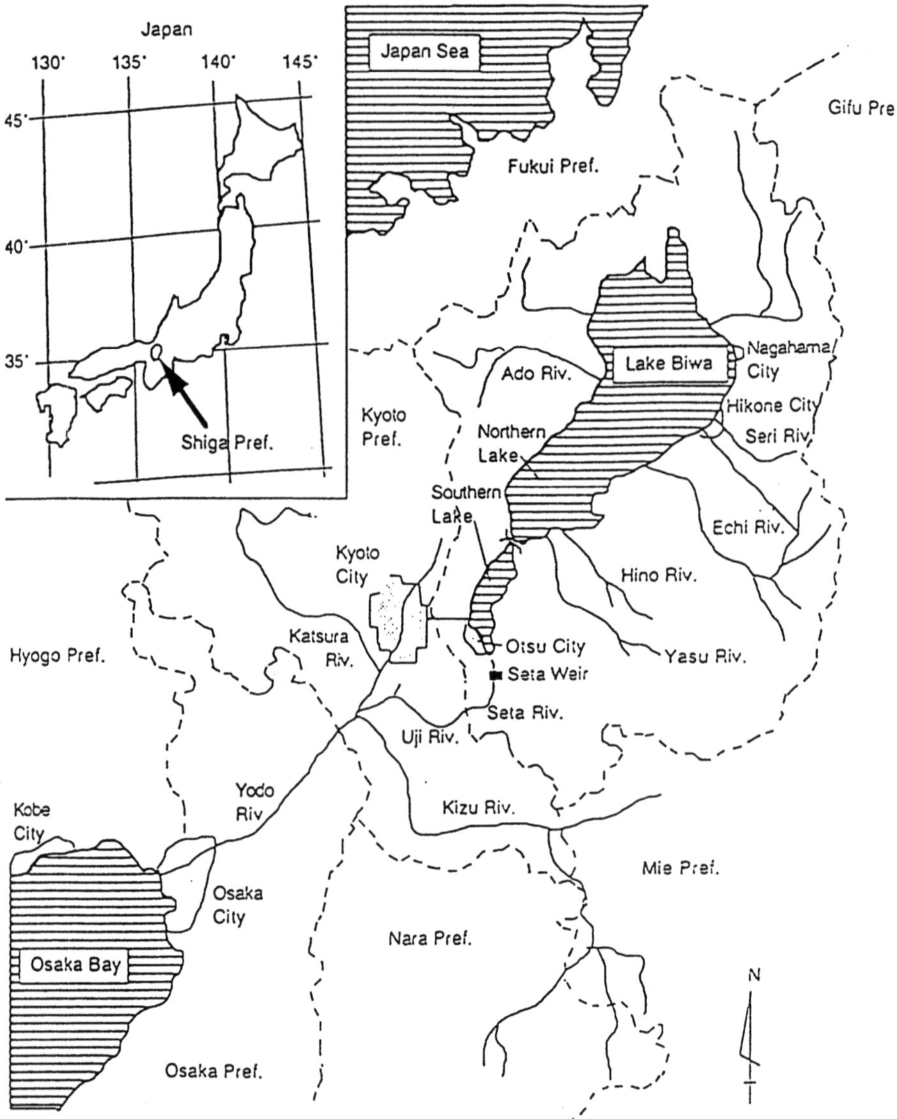

Source
'Lake Biwa', Shiga Prefectural Government, 1991

Figure 3
Changes in water level of Lake Biwa

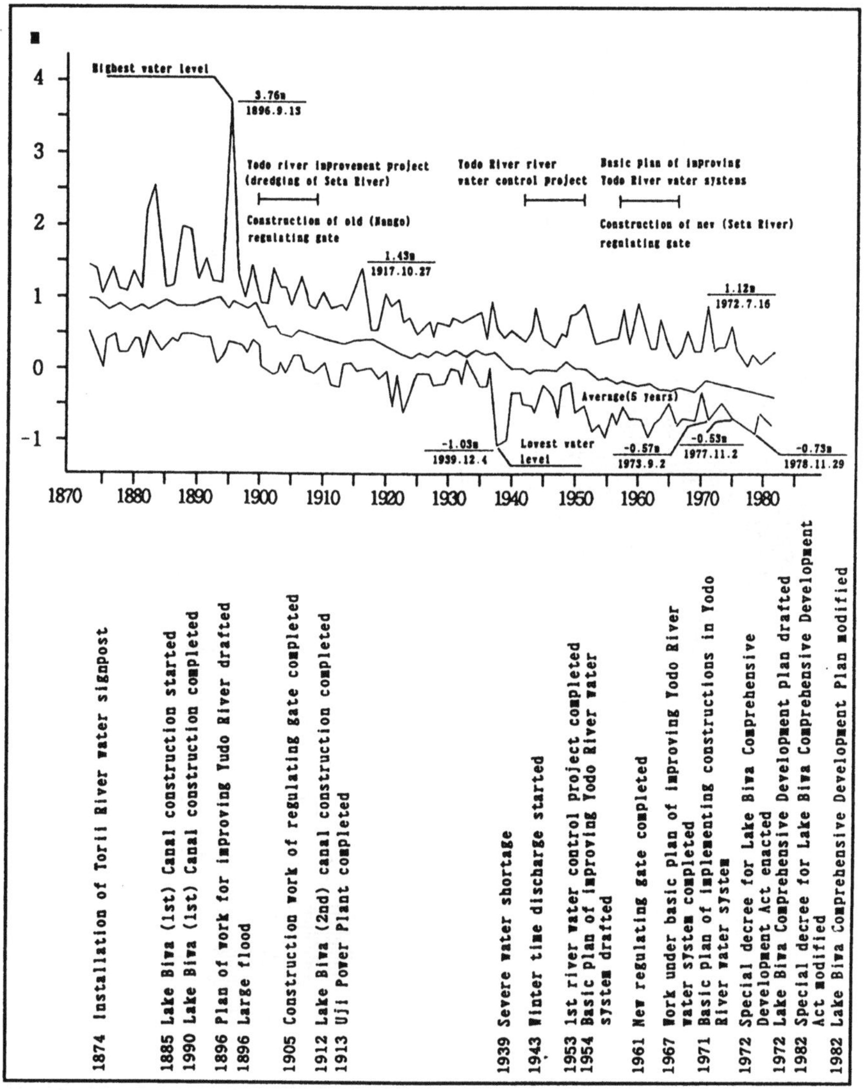

Source
'Lake Biwa', a brochure published by Kinki Regional Construction Bureau, Ministry of Construction, Japan

Figure 4
Topography in the Lake Biwa Catchment Area

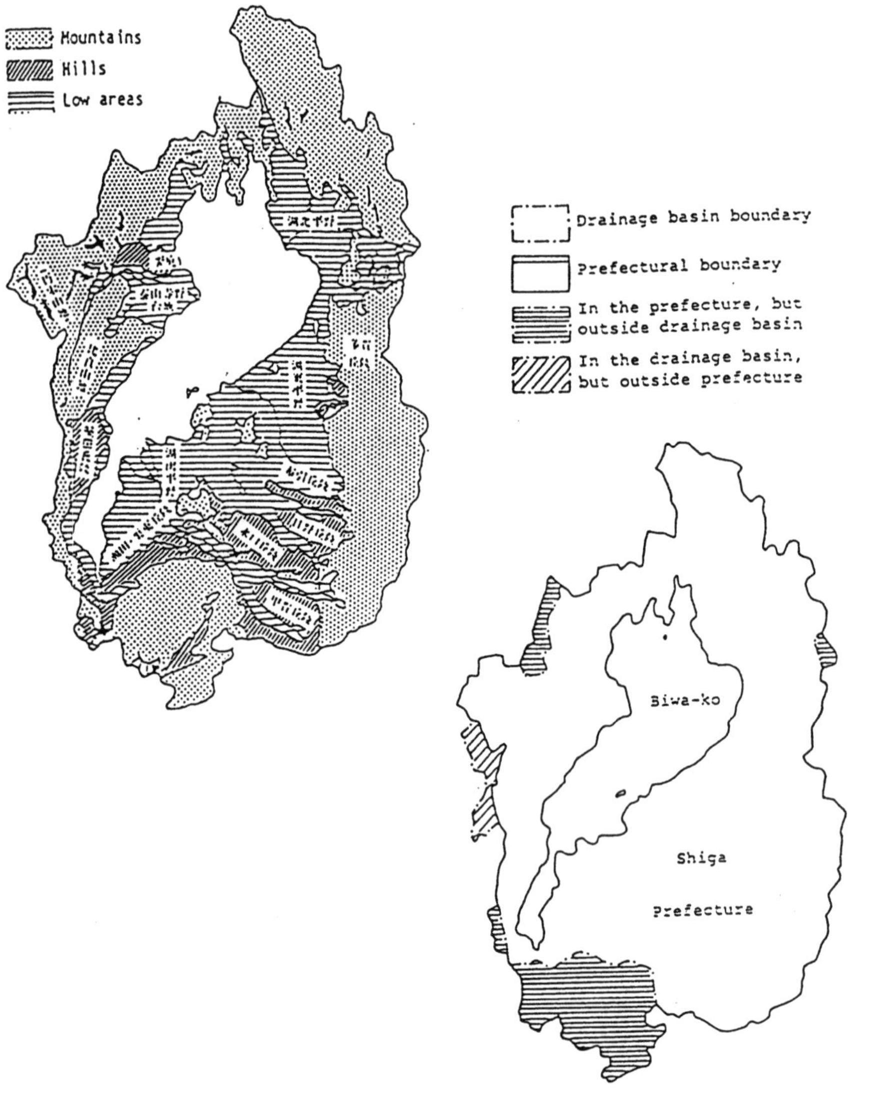

*Comparison of the boundaries of Shiga Prefecture and L. Biwa's drainage basin.

Source
Databook of World Lakes, LECS '84, Shiga

Figure 5
Cause-effect diagram on pollution, public reaction and local government responses

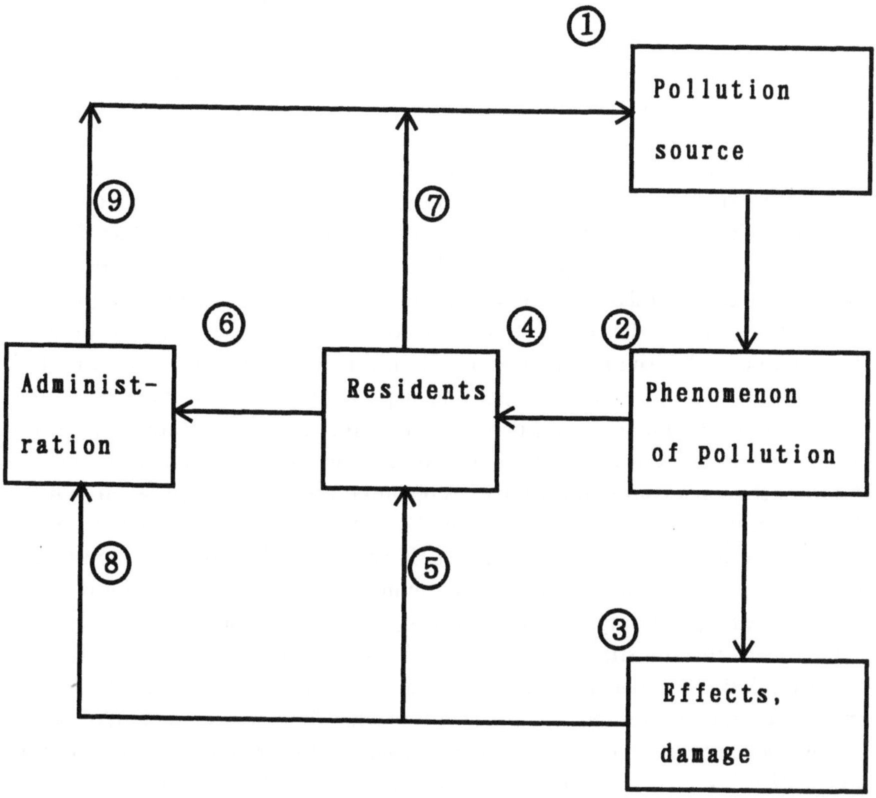

IV.5 Remarks

Dr Salim

The discussions on technology showed the direction of present efforts toward resource management and pollution management. On resource management we are informed of various approaches including the future clean energy technology, the photo-type efficiency system and the pollution impact management in relation to the disposal of CO_2 in the ocean. Case studies from Thailand, Germany, Kawasaki, and Lake Biwa amply indicated the development of the peoples' participation. On global cooperation, we notice that there are concrete proposals like the global energy infrastructure cooperation in the Asia-Pacific.

A series of presentations prove that the environment is being degraded, and therefore activities related to the control of climate change become important. It is in that context that energy issues become a high priority. A final point is that we notice that the research proposals and research activities concentrated on company sources, such as a space solar power system and dumping in the sea. Both space and sea are a common resource, however. Here, a question arises. To what extent can research technology development take place where the impact is on a common resource? A resource is not owned by Japan, but by the human beings on the globe at large. Therefore, it seems to be necessary to set in place a kind of code of conduct. If the research may have significant impact on the common resource, it may be necessary to inform the world about its possible negative impact before moving toward its implementation.

That leads to a question, a moral issue. To what extent are the researchers in Japan guided by responsibilities across the globe if the research may have an impact on a globe common resource? This is what the Science Council of Japan should look into in order that the world at large will feel more comfortable in seeing that whatever is being done here in Japan will not negatively affect global common resources. Therefore those engaged in research and technology have a high responsibility, I may say a high moral responsibility, that whatever the outcome of the research it should not be negatively affecting global common resources.

Dr Krupp

In 1980 technology assessment was officially stopped in this country. I've seen the MITI files and they ended in a final volume, eight centimetres thick. I think such assessment and self-assessment are necessary especially from institutions and organizations of larger projects. The United States has been exercising this sort of thing for 20 years and other countries have followed suit. So why not propose technology assessment as a kind of must? There are various ways of doing this, but basically it should be permanent and ongoing. When new results become available, they would easily be fed back into the technology assessment machinery whether or not they throw new light on results obtained earlier. So it is a self-correcting recursive process.

The cost benefits in these methods are not easily balanced and how can you quantify them? But still I think that's one of the first steps toward having some accountable discussions. My point is that the issues surrounding the global environment are not necessarily the same as those involved in local environment. I think that the viewpoint from the grassroots and, say, from the satellite, are two points that should be combined. The conflict of interest is not merely between the developing and the developed countries but between the present

IV.6 Session Discussions

Professor Sawa

Dr Salim talked about the need for technological assessment on the dumping of CO_2. This can give rise to a new type of environmental issue to which we would have to address the question of economics. Of course, we cannot calculate the economic situation 100 years from now. But an opening should be made for the technological feasibilities, and if they don't work, we should abandon them. Then we will have to talk about the physical energy balance or psycho-assessment. In other words, a so called objective assessment is required and, if the results seem feasible, we should go further. Given that the energy system requires so much time for conversion it has to be on a very extensive scale to be effective and efficient. Therefore, the so called cost capacity curve must be considered, otherwise we cannot make wise judgment about what will happen in 100 years.

Dr Cho

In my paper I emphasized the role of government or official meetings among countries in this region. I also touched on the important role that NGOs can play. But I think those two are not contradictory. They are complementary. But now the problem is this. We are facing a changing situation because of the conflicts of interest among countries in this region. That is beyond the control of NGOs. I emphasized the role of government commitment here.

Chairman

Thank you. Now Mr Nakamura.

Dr Nakamura

I agree with Dr Salim about the need to promote a global sense of involvement. But there are a number of very specific obstacles that we have not quite been able to define. They may have a lot to do with the current mechanisms of trade as well as market orientation. But it may also have more to do with structural impediments, what I call the environmental structural impediments that undoubtedly exist between the developing countries and the developed countries.

Chapter Five

New Initiatives for Environmental Protection

V.1 Overview

Chapter five deals with the issue of new initiatives for environmental protection and contains two papers. It is in this chapter that summaries are made after the lengthy discussions.

The first paper by H. Krupp discusses the need for concrete and feasible action programs for each problem raised. At the same time, he argues that what is needed is transformation, not just in our ways of thinking about environmental protection, but in our ways of living. This is an issue, he says, for every citizen on the globe. Without such massive support toward the protection of our globe, a mountain of paper work for sustainable development or environmental protection would be forgotten and have no visible results at all.

T. Sawa, on the other hand, critically scrutinizes in his paper the recent history of environmental protection which is not free from the so called 'tri-dilemmas'. Among others, he points out the dilemma of development and environmental protection. With a rapidly developing technology, however, he believes many things can be done. Yet, he argues, those efforts would not be enough to free all of us from the tri-dilemmas mentioned above. In this regard, a combined search for energy efficiency, recycling, and kind dumping—plus technological breakthrough—should be made through international, national and local cooperations.

V.2 Summary Speech

H. Krupp
Germany

Abstract

As requested by the organizers, this paper presents an overview of the conference and my commentaries. This cannot be anything but selective and subjective.

In my opinion, the papers presented at this conference may be classified into three categories:

- technology-oriented papers which review new technologies claimed to be able to ameliorate some of the ecological problems encountered;
- political papers, ranging from local to federal and global levels, which present the instruments, dilemmas and hopes of environment-related politics;
- papers from the social sciences which stress how inextricably technical and ecological problems are embedded in society.

This paper:

- highlights major points of agreement and doubt;
- presents clarifying quantitative data;
- outlines a societal theory (Schumpeter Dynamics) to account for the phenomena observed;
- tries to draw conclusions about possible action to be taken and distinguishes between plausible and less plausible projected futures.

Agreements and Doubts

From the papers of this volume I conclude that there is broad agreement on the following two points:

(1) There is a wide range of technologies available, both for increasing energy productivity (efficiency) and for introducing renewable energy sources. They are in different stages of development, ranging from laboratory design to demonstrated effectiveness. In particular, the paper of Mark Levine suggests

that, by and large, the available technology may double energy productivity (GNP per unit of energy) at no or even negative costs within a reasonable time span. Therefore, a simplifying slogan says that, within the next decades, an increase in energy productivity will be the cheapest source of energy.

(2) As for renewable energy, wind energy seems to be ready for diffusion on a larger scale. As shown by Koichi Yamada, within the next one or two decades photovoltaics may follow suit. Thermal solar energy is already widely being used and its diffusion continues. But, both wind and solar energy are fluctuating sources which need to be complemented by energy storage and/or to be integrated in a multiple source electricity grid. Major control problems may still have to be solved. Apparently there is no principal stumbling block to the economical application of these technologies on a large scale in the next century.

However, there continues to be fundamental disagreement on nuclear energy. As a consequence, Takao Kashiwagi's report on projects of nuclear fission and fusion met with unabated scepticism. This seems to be stronger and more outspoken in Europe and in the United States than in Japan. The whole nuclear cycle, from mining to the final disposal of the burnt-out fuel, can hardly be claimed to be 'sustainable', however undefinable this term may be.

Also, satellite power stations are not being seriously considered in countries other than Japan. Cost, atmospheric disturbances and danger to people are major counter-arguments.

Although discussed only briefly during this conference, there is a major difference between Japan and all other OECD countries as regards technology assessment. This procedure of carrying out a broadly based, science led debate on the societal acceptability of major technologies is more or less standard practice in the west, but apparently not so in Japan. For example, CO_2 dumping into the oceans, discussed by Hisashi Ishitani and Ryuji Matuhashi, seems to require technology assessment because of unknown and untestable chemical reactions and long-term oceanic circulations. Similarly, the recycling of CO_2 in methanol production, proposed by Koichi Yamada, ought to be reassessed in the light of alternatives, such as the avoidance of the carbon cycle and the exclusive use of renewable energy, because the large-scale collection, transport, and processing of CO_2 is very eco-intensive.

But at the same time, there is universal scepticism that technology assessment can lead to a major redirection of technological evolution. The reason is the anomy of modern societies, as shown below.

As for environmental stress resulting from the use of materials and the production of waste, there is agreement that recycling, re-processing and new design may bring about some improvement. Thus, taking the potential increase in energy productivity as a proxy, it may be assumed that, worldwide, the per capita rate of energy and materials as well as the per capita production of waste might be halved. On the other hand, however, a complete decoupling of resource use, environmental stress and economic growth seems to be out of sight, if not impossible.

The Widening Gap between Resource Productivity and Economic Growth

As already indicated, resource productiviy is defined as the ratio between economic growth and the amount of a particular resource consumed. In the wealthy countries, the productivity of energy and other natural resources increases by about 1% per year. In developing countries, however, the productivity tends to decrease; it increases only in later stages of development. With an optimistic long-term perspective and for the sake of simplifying the argument, let us assume that the productivity of the use of energy and materials can be continually increased by about 1% per year.

On the other hand, the rate of using resources and the rate of waste production are programmed for major increases. The principal factors are:

- increase of population by a factor of two or more until the middle of the next century;
- further economic growth of the OECD world by probably at least 1% per year, on average;
- much larger economic growth in major parts of the developing world, in particular in population rich countries.

As a result, the projected rates of increase in resource productivity are bound to be greatly over-compensated for by economic growth. Environmental stress may therefore grow by a factor of 10 in the next century.

To pick just one example, increases in the fuel efficiency of automobiles are being over-compensated for by increases in fleet size. Presently, fuel efficiency is even decreasing in many countries because bigger and more powerful cars are being used and because of more careless driving habits. In Japan, this turn of trends occurred in 1982 and continues today. Extrapolating from the situation in the OECD countries, in the next century the global fleet of passenger cars might grow to a few billion, if self-suffocation of individual transport in the mega-cities does not counteract these trends.

Several authors have shown that, as a result, the present resource consumption of the OECD-type countries would have to be reduced to one tenth or less in order to limit the growing environmental destruction to 'tolerable' proportions.[1]

The many anecdotes about local successes in the abatement of one type of environmental stress or the other, as mentioned for example by Lester Brown, cannot be generalized to indicate a global trend. All sufficiently qualified and quantified government reports on the state of national environments, as well as the publications of the Worldwatch Institute, provide ample evidence of an overall worsening trend.

The wide range of available policies and measures to be taken, particularly discussed by Rolf-Ulrich Sprenger, might alleviate the problems, but:

- probably only temporarily, because the creativity of growth-oriented innovation would tend to circumvent them. As a result, a race between regulatory and growth-oriented innovation might ensue. This occurs, for example, in the control of bio-cides in agriculture, of chemicals in wood preservation and of additives in food;
- their enactment is slow, or non-existent. Dramatic examples can be drawn from all countries. In Germany, the pledge for an abatement of CO_2 emissions by 25% until 2005 is not being fulfilled by the German government. The present reduction by 11% is due to be the breakdown of the east German industrial base, the slackening of the economy and two exceptionally mild winters;
- insuperable problems would probably be faced if ceilings and charges in emission control had to be quantified in an international framework, including the poor countries.

Fast progress with regard to environmental politics, predicted for the near future by Lester Brown, does not appear to be in sight.

Frustrations

The above results frustrate citizen groups, lawmakers, scientists and the people at large affected by environmental degradation. Various examples have been discussed by Suraphol Sudara, Rong-Hui Huang and Masahisa Nakamura at this conference. As a result, the debate turns to scapegoating and governments, political leaders, business and the greed of consumers are identified as culprits of the self-destructive trends. Specifically, the poor world accuses the wealthy countries of a lack of consideration of its difficulties and of selfish trade and financing practices, as forcefully brought forward by Emil Salim at this conference.

Another approach in the OECD world is to moralize, resulting in appeals such as 'we must change life-styles'. So far, this has not been particularly successful.

Yet another approach, predominantly exercised at this conference, is the insinuation that somehow the 'trilemma of resources, economic growth and environment'—a term introduced by Mark Levine—can be solved without sacrificing growth. As shown above, this does not seem to be the case.

Some countries even claim to have solved the trilemma. In the case of Japan this is done in the paper of Saburo Kato. This can however neither be confirmed nor generalized.[2]

It is a general dilemma in this world that daily action is dominated by short-term considerations. In economic terms, expectations of short-term wealth tend to supersede considerations of long-term welfare. The reasons are discussed below.

A World of Explosions

Why is it that there seems to be no convincing evidence that politically and within society the trilemma resources/economic growth/environment can be solved? Before outlining a model to answer this question, I would like to draw your attention to the fact that we seem to live in a world of explosions.

Commonly, the growth of the global population is termed 'population explosion'. The present annual growth rate is somewhere between 1.5 and 2%, and our world is not only experiencing a population explosion, but also explosions of economic growth, use of energy and other material resources, crime, mass migration, implosion of mega-cities and efficiency of weapons,[3] and perhaps others too. Geographically, the intensities and growth rate of these effects are of course quite inhomogeneously distributed over the earth.

Other than their exponential character, these trends show symptoms of global convergence. Despite time lags and different starting levels of these effects in different countries, global commonalities tend to dominate over cultural differences. Yet major differences still prevail, for instance with respect to societal cohesion, educational performance and crime.[4] But, in the long run, growing similarities in technology, economic structure and consumption patterns may win against residual cultural diversity. We shall all equally be in the same boat: spaceship earth.

In the face of all these explosions and the poverty prevailing in large parts of the world, it does not come as a surprise that environmental destruction is not foremost on the agenda of most people. Apparently, there are much more pressing needs on people's minds.

In the next section, a pertinent model is proposed.

Schumpeter Dynamics: a Proposed Model of Society

The model's focal points are:

- modern societies have become differentiated into societal subsystems;[5]
- at the same time, we observe their re-integration into what I call the system of Schumpeter Dynamics through structural coupling;
- the evolution of Schumpeter Dynamics is being 'dynamized' by technological and societal innovation;
- because of widening and intensifying networking, Schumpeter Dynamics is 'globalizing'.

The main particular subsystems relevant to this study are the economy, the political administration and parties (politics), technology and the world of consumption. Their main characteristic is their autonomy, determined and represented by their particular codes of communication:

- The economy deals with monetary transactions, accounting for profits or losses, growth or decline, or related binary codes. The relative autonomy of economic communication leads to externalities, such as waste deposited outside the system, into the environment, for that matter.

- Politics communicate in terms of power: government or opposition, increase or decrease in the electorate, in party strength etc. An extreme externality of politics may be the waging of a nationalistic war.
- Technology also constitutes a subsystem of its own in society. Its code of communication is the increase or not in technological know-how. Externalities may be caused by technologies such as nuclear energy production. From within the system there are calculated risks incurred by technological advance. Seen from the outside, they expose the general population to potentially disastrous danger.
- The world of consumption has its own semantics: to buy or not to buy, in or out of fashion and what have you. An external cost of consumption is environmental damage.

If this model of societal differentiation holds, modern societies lack central control. They are made up of subsystems with autonomous communication relative to which centralizing forces, such as universal ethics, are getting lost. Societal cohesion tends to dwindle. The result is anomy.

The inherent instability is at least partially compensated by re-integration into Schumpeter Dynamics through structural coupling between these subsystems. Examples for this inter-systemic structural coupling are:

- political stability tends to grow with economic growth;
- the political administration provides the economy with an infrastructure of traffic/transport, medicine, education, research etc and relieves it from having to supply social services, in particular the support of the unemployed, sick and aged;
- technology provides innovation which, in turn, strengthens the economy and the political administration;
- consumers enjoy increasing wealth from this interplay and support it on commercial and political markets.

Through their structural coupling, the four subsystems constitute a synergism which is driven by competitive innovation and growth.

As indicated, Schumpeter Dynamics leads to international networking. Mechanisms include technology transfer, trade, banking, trans-national corporations, development aid, GATT, regional pacts and so on. In particular, after the breakdown of the Soviet world, there does not appear to be any major resistance to this global spreading of Schumpeter Dynamics.[6]

In summary, if this proposed model of Schumpeter Dynamics is valid, it stipulates that:

- global destruction is an externality of the selfish subsystems;
- there is no integrating center of (self-)control;
- in view of drastic geographic and social disparities, global solidarity is improbable;
- development occurs by evolution.

This is corroborated by a look at the main preoccupations of OECD-type and other countries. In the media, in election campaigns and in the policy statements of the major organisations and institutions, the following subjects prevail:

deregulation, removal of international trade barriers, 'pump priming' in order to stimulate the economy, lowering of taxes and interest rates, formation of regional blocs, globalization[7] and military interventions in non-aligning countries. These are all aimed at stimulating Schumpeter Dynamics.

In contrast to this, there are only minor attempts to intervene on behalf of the environment. Major attempts, such as the introduction of a substantial CO_2 or energy tax, have been defeated so far, both in the European Community and in the United States. They have not yet even been attempted elsewhere.

This is a more reliable indicator of world trends than widespread self-rewarding eco-rhetoric.

Action

Essentially, Schumpeter Dynamics is an evolutionary model. It maintains that world dynamics are enacted by supra-individual forces. On the other hand, it is of course people's actions which drive the world. Business leaders and politicians try to develop strategies and to make decisions in order to implement them. Individuals make choices between alternatives and perceive individual freedom. How can this be made compatible with the workings of a systemic dynamism?

People's freedom of choice is not absolute. It is pre-structured. This is the principal reason why people have to be schizophrenic with regard to environmental protection and pollution: they more or less have to own an automobile, they all use energy which is produced in polluting power plants and so on. Business leaders are in competition with other companies and have to fight for profit margins. They cannot worry too much about the long-term benefits to the general welfare of the world. Politicians must adhere to a narrow range of rhetoric and action to stay in power. All this reduces everybody's range of alternatives drastically, whatever their individual convictions might want them to do. They can only choose between loyalty and exit.[8] In the latter case they jeopardize the financial situation of their family and their staff. In other words, only actions which are in resonance with the evolution of Schumpeter Dynamics are successful and do not become damped out by the system.

But the next question is, how is innovation possible? As an example, how could the ecological movement arise? Innovation means changed or new structures, institutions, rhetoric and so on. The resonance spectrum of Schumpeter Dynamics has a dynamic profile. One of its new resonance peaks, however low, is ecological communication.[9]

Under the impression of ecological catastrophes, scientific evidence and media coverage a new system of communication has been formed which is penetrating Schumpeter Dynamics.

- Through structural coupling in the business world, it has generated immune reactions, such as negation of evidence, suppression of its diffusion and negative sanctioning of proponents of ecological communication. This has been a first step. The next one has been 'greenwashing', whereby the products and transactions of companies have been given a new, more environment-friendly profile. Corporate identities have been tinted green and eco-balance sheets are now being drawn up. Finally, eco-business has ensued, offering many new products claimed to be environment-friendly, healthy and safe. At the same time, the programs of research and development which aim at more efficient use of resources and at renewables have been strengthened.

- After the emergence of citizen movements and, finally, green parties in some countries, the political rhetoric has started to include ecological principles in prominent positions of party programs and election campaigns. At the same time, environment-friendly re-regulation is proceeding, although in small steps.
- Technology is responding by offering new options. Of course, all options are ambivalent because the manufacturing, and use of new environment-friendly equipment is also using up ecological capacity. Waste incineration, for example, pollutes the air with highly poisonous dioxins. Some proposed photovoltaic materials to be spread over large areas are toxic when released from the panels by erosion. Material reycling requires chemical plants which are not emission-free. Mass genetic engineering is probably the most risky new technology because its products cannot again be removed from the environment and start irreversible vectors of evolution.
- People respond by sorting garbage and choosing different products, but life-styles and mass consumption behavior remain basically unchanged.

In the light of these observations I cannot share often propagated visions of an imminent ecological revolution. Instead, I see only a gradual evolution at rates to be measured in units of human generations.[10]

In conclusion, the evolution of Schumpeter Dynamics is the framework which shapes human action. Permanent societal fluctuations and human spontaneity are being filtered accordingly. Of course, at the same time they germinate the transformation of Schumpeter Dynamics.

Transformation of Schumpeter Dynamics

However high the accompanying noise levels may have been, initiators and promotional carriers of the transformation of Schumpeter Dynamics have been citizen groups, green parties, ecological research institutions, particular 'green' media and national and international networks of 'green organizations'. In later stages, these mostly non-governmental organizations (NGOs) started to include business initiatives which blurr their targets.

In addition, there are a large number of environment-oriented national and international networks, including even the United Nations. The two conferences, 1972 in Stockholm and 1992 in Rio de Janeiro, testify to this. However, at the same time, these have been events to voice the desires of the developing world for economic growth and international aid.[11] This is in agreement with the proposed model of Schumpeter Dynamics which stipulates that each subsystem of Schumpeter Dynamics interprets and communicates external signals in its own terms and reshapes them accordingly. For the developing countries, the ecological plight constitutes a unique chance to economize on its ecological wealth and to stress their call for financial transfers from the rich.

In co-evolution with the industrial revolution, originally hierarchical societies have become restructured. As a result, a compartmentalized society has emerged. This evolution does not appear to be reversible.[12] In the resulting absence of control, the future transformation of societies would have to occur from within by the interplay of all societal forces. This will be a tedious process of trial and error. All the time, both from advocates of a conservation of Schumpeter Dynamics and from 'green revolutionaries' fighting for its transformation, a great number of paradigms are being launched, diffused by the media and reshaped into the particular communication within the different subsystem. It is the resulting cacophonic communication processing which drives evolution forward. In contrast, residues of former societal models, based for instance on non-partisan leadership by a bureaucratic elite of civil servants, will probably not survive. No other elite seems to be able to take its place, whether scientific, political, charismatic or military.

If this is so, the interplay of all societal forces within the usual limits of liberal laws affect societal evolution and may cause a transformation of Schumpeter Dynamics.

If this is accepted, it follows that it would be systemically rational to promote citizen movements and their juristic instruments, to facilitate the access of green parties to parliaments at all political levels (municipal, prefectural, federal) and to strengthen democratic participation in all institutions, public and private. Freedom of information would be a prime requisite.

Obviously, with respect to ecological concerns, such a liberalization works both ways. It also strengthens the voices of the world of consumption and the economy which are self-centered and would not easily agree to belt tightening.

Therefore, it is my conjecture that because of the inertia of Schumpeter Dynamics, including that of the world of consumption, much more intense ecological pressure will be needed to trigger a transformation of Schumpeter Dynamics more radical than at present. The transformation have would have to affect all parts of society deeply.

The socio-psychological essence of Schumpeter Dynamics, its 'ethics', is individualistic. Therefore, one of the principal paradigms of our times is de-regulation in order to assure the greatest possible freedom for entrepreneurial and consumer action.

However, the threat to the global commons cannot be alleviated without networks of negotiations arriving at compromises, including even the poor countries. They would have to entail major financial transfers from North to South in exchange for appropriate resource management in the South. Another feature might be joint projects so that, for example, solar energy in the sun-rich belt could be captured and transported to the North in various forms, particularly electricity, hydrogen and/or methanol (see Koichi Yamada).

The principal requirement would not only be a reduction in growth, but a basic change of life-styles, as a widespread slogan says. Mass consumption at today's level would not continue to be possible. In other words, it would have to be a world quite different from that of today.

Late in the next century, this may entail a bifurcation:

- self-transformation of Schumpeter Dynamics to achieve, say, 10 times as much resource productivity and a drastic reduction in growth. This is unimaginable from present OECD viewpoints; alternatively,
- gradual catastrophic self-destruction of Schumpeter Dynamics to be followed by a new beginning. Possibly a less tightly coupled, less population-rich and more diversified world might then become more stable. As is mass consump-tion, population density is a key variable which determines global evolution.

This is highly futuristic, but so is the long-term future.

From a theoretical viewpoint, another bifurcation might be:

- continued self-abandonment to evolution; or
- reflective self-control.

The latter cannot be excluded, however improbable it may appear from our historical experience up to today. Deregulation would have to give way to dense networks of re-regulation, whether achieved by evolution or self-control.

Conclusion

I can only fully agree with Takamitsu Sawa who calls for a 're-examination of the industrial civilization of the 20th century based on mass production, mass consumption and mass waste' and warns of the 'pitfalls of the belief in the omnipotence of technology'.

In natural history, many species evolved into a dead-end and became extinct. However, humankind is the most flexible of all. Self-rescue cannot be excluded as a possibility. Incrementally positive factors have been described above, however massive the inertia of Schumpeter Dynamics may be. But unpredictable events, such as human migration, drastic population shrinkage and natural or technical catastrophes may create entirely new boundary conditions so that the future continues to be wide open.

For a comfortable future, Schumpeter Dynamics seems to enable too little too late.

I most gratefully acknowledge the assistance of Frau Linhart-Bach to improve my non-native English.

Footnotes

1 Ohno, Yotaro: 'Greenhouse gas emission, control and sustainable development', Energy in Japan, number 110, 1991, pages 24-33, as well as Weterings, R. A. P. M., und Opschoor, J. B.: 'The ecocapacity as a challenge to technological development',

manuscript submitted for a workshop of the Dutch Committee for Long-Term Environmental Policy, March 22-24, 1993 at Zeist.

2 As for international comparisons, Takamitsu Sawa makes the recommendation that data on CO_2 emissions or primary energy use, per unit of GNP, ought to be recalculated on a per capita basis in order not to be misled by the fluctuation of currencies. An example is the large discrepancy between the domestic buying power of the yen compared with its foreign exchange rate in the West—possibly a factor of two at the time of this conference. On this basis, causative factors of environmental stress are more directly related to the consumption patterns of the people.

3 The cost of killing a person by bombing Dresden (Germany) in 1945 was 10^5 times higher than it would have been in a possible biological warfare.

4 The annual numbers of enforced robberies in Tokyo, Frankfurt/Main (Germany) and Chicago differ by factors of 1, 40, and 200, respectively.

5 This model is based on the fundamental and extensive analysis of Niklas Luhmann. His theoretical base has been summarized in particular in his article in Cardozo Law Review, Operational closure and structural coupling, volume 3, number 5, 1992, pages 1419-1441

6 Mark Levine raised the relevant question: what is the difference between this model and capitalism? It is structural. Capitalism tends to be discussed in terms of a stratified model in which the economy imposes itself on politics and the world of consumption. The structural coupling between the economy and politics described in 'The industrial state' (Galbraith) is a step in the direction of Schumpeter Dynamics.

The difference is also theoretical because modern system theory permits a much more systemic elaboration of societal interactions than classical analyses of capitalism.

7 These figure prominently also in the enunications of the Hosokawa Government.

8 Hirschmann, Albert O.: Exit, voice, and loyalty. Harvard University Press Cambridge, Massachusetts 1970

9 Luhmann, Niklas: Okologische Kommunikation. Westdeutscher Verlag Opladen, 3rd edition 1990

10 See my papers at the preceding IBM Japan International Symposium Energy and Environment, October 21 to 24, 1990, in Takeuchi, Kei, and Masatoshi Yoshino (eds) The global environment. Springer Verlag 1990, pages 162-170

11 The trilemma-generated dilemma of UNECD in Rio de Janiero in 1992 is analogous to the national situation in Japan as described by Tadashi Kawata. In 1967 a fundamental law on environmental protection was passed by the Japanese diet which contained a clause that economic growth must, however, not be impaired. Similarly, UNECD has postulated 'sustainable' growth. In Japan in 1970 this clause was dropped, although growth continues to dominate political and economic action.

12 Except under extraordinary stress. Examples are nationalistic rallying in cases of war (recently in the Falkland Islands and in Iraq). Environmental catastrophes in the future may work similarly.

V.3 Appraisal Speech

T. Sawa

Director and Professor, Institute of Economic Research, Kyoto University, Japan

Abstract

This paper outlines ways to deal with the trilemma of environment, energy and growth. It looks at rates of the world's energy consumption in the past 20 years and the resurgence of interest in the problems of the global environment.

The paper suggest that industrialized nations must extend their time horizon to 50–100 years and globalize their perspective. It looks at energy alternatives and questions the belief in the omnipotence of technology. The paper outlines the need to re-examine the industrial civilization which is based on mass production, mass consumption and mass waste, and move from an 'industrial' to a 'metabolism' civilization.

1 Energy and Environmental Questions: Why Are They Raised Now?

Toward the end of the 1960s, there developed in the industrially advanced nations of the West and in Japan a sense of criticism and reflection over the prevailing obsession with economic growth and the perilous destruction of the environment. Industrial pollution (such as air pollution and water contamination) and urban pollution were on the increase, as was the occurrence of pollution related diseases, the most notable here being Minamata disease.

The Limits to Growth (1972), a study commissioned by the Club of Rome, sounded the following alarm: 'If population growth, industrial production, environmental pollution, and consumption of food and natural resources continue at current rates, the growth of the world economy will reach a limit within 100 years, resulting in the collapse of the world order. The only feasible option for averting this disaster is to revise the existing growth pattern and make a heroic transition to static equilibrium, that is, zero growth.'

Subsequently, the first oil crisis of 1973 and the second crisis of 1979 not only led to sharp rises in oil prices, but also revealed the scenario of depleted petrochemical resources. As a result, the world was forced to pay closer attention to the energy problem. The decline in crude oil prices since 1986, however, significantly averted our attention from the problem. So much so that by the end of the 1980s the term 'energy saving' became almost obsolete.

The changes in the gross national product (GNP) elasticity of energy consumption (which is defined as the percentage change in energy consumption divided by the percentage change in real GNP) have been as follows: 1.28 (= 11.8/9.2) between 1965 and 1973; 0.23 (= 0.9/3.9) between 1973 and 1979; -0.11 (= -0.4/3.6) between 1979 and 1986; and 0.85 (= 4.1/4.8) between 1986 and 1991. The protracted economic slump since 1991 has resulted in the diminishing rate of increase in energy consumption and a reduced volume of CO_2 emissions.

The heightening interest in the problems of the global environment since the summer of 1988, particularly with the problem of global warming caused by CO_2, rekindled world interest in the energy problem. In fact, the issue was placed for the first time on the agenda of the June 1988 meeting of the 'group of seven' (the major industrial powers) held in Toronto, Canada. At the Paris/Arche Summit of 1989, global environmental issues occupied one third of its economic communique.

The cause of this resurgence of interest was the end of the cold war. The problem of the global environment became identified as a problem common to all humanity in the post-cold war era. In order to curtail the emissions of CO_2, the consumption of fossil fuels has to be reduced to the greatest extent possible. How can the consumption of fossil fuels be curtailed?

It is necessary for us to extend our time horizon to 50-100 years and globalize our perspective. If we face up to the reality of the North/South problem and the problem of population explosion, we can clearly identify the environmental and energy constraints and the structure of the trilemma of environment, energy and growth.

2 Problem of the Global Environment: What is its Nature?

What is being asked is a re-examination of the 'industrial civilization of the twentieth century' based on mass production, mass consumption and mass waste. We must explore the feasible direction of 'sustainable growth' under resource and environmental constraints. The totality of macro-economic policy, business behavior and consumer behavior has to be reassessed.

Since the latter part of the 1980s, the life-style of Japanese consumers has become excessively energy dependent. The prevalent aesthetics is that extravagance is 'in'. This extravagance orientation has to be transformed into a preference for a simple life-style.

What about the prospects for nuclear power generation? Although nuclear power generation is free of CO_2 emissions, it faces the difficult problem of public acceptance of plant site location. The accident in April 1986 at the

Chernobyl nuclear power station in the former Soviet Union had a serious adverse effect on the public's acceptance of having a nuclear reactor in its midst. How can public acceptance be secured? A hard-headed but soft-hearted approach to the energy policy is called for.

What is the optimum mix of energy sources? According to the Japanese government's *Long-Term Outlook for Energy Supply and Demand,* actual nuclear energy generating capacity in 1989 was 28.9 million kW, or 9% of the country's total energy supply. The projection for the year 2000 is 50.5 million kW or 13.2% of the total, and 72.5 million kW or 16.7% for 2010.

What about the prospects for renewable energy such as solar energy? At present, solar batteries are too expensive to be economical. Incentive measures are needed to promote their mass use. A substantial cost reduction is expected to derive from the economics of scale from mass production. According to the *Long-Term Outlook for Energy Supply and Demand,* the percentage share of new energy sources, including solar energy, of the total energy supply was 1.8% in 1989, and is expected to rise to 2.9% in the year 2000 and 5.2% in 2010.

3 Science and Technology Versus Political Economy

I would now like to discuss the pitfall of the belief in the omnipotence of technology, as suggested by Malthus in his *Principle of Population* (1798) and in the *The Limits to Growth* (1972) of the Club of Rome.

What can science and technology accomplish? Think of the paradox of CO_2 balance; the need for 'economic methods' for promoting technological development and its practical use; and issues concerning the carbon tax.

What is the nature of the civilization of the twenty-first century that will replace the industrial civilization of the twentieth century? It requires a transformation from the 'throughput' civilization to the 'metabolism' civilization. The result is a civilization that incorporates optimum consumption, minimum waste, recycling, energy saving and longer product life. An index of environmental health, or 'green GNP', replacing the conventional GNP is being sought.

The energy dependent society par excellence is the United States. Vice President Gore, among others, has proposed the setting of targets for CO_2 emissions, a fundamental shift in the pattern of transportation, replacing incandescent lamps by fluorescent lamps, 'down-sizing' automobiles, improving automobile fuel efficiency, raising electric utility rates and promoting the wider use of natural gas.

For reference, the energy requirements for passenger and freight transportation are compared below.

(a) Passenger transportation

Category	Energy required
Passenger cars	510(kcal/passenger-mile)
Railroad	47
Airplane	529

(b) Freight transportation

Category	Energy required
Trucks	923 (kcal/ton-kilometer)
Railroad	60
Airplane	5,249

4 Energy Saving and Technological Development: How can a Transformation of Civilization be Achieved?

Energy saving and technological development require economic incentives. An effective means for effecting a 'transformation of civilization' is to appeal to the aesthetics of consumers. Consumers should count 'eco-friendliness' of products as an important purchase quality.

Japan plays a crucial role in the global efforts to tackle environmental problems, and should be fully aware of this responsibility.

If the industrially advanced countries, including Japan, want to hold the level of CO_2 emissions at the 1990 level, while recognizing the 'right to growth' of developing countries, they must reduce the consumption of fossil fuels absolutely, not merely reduce the rate of increase in their consumption. To that end, they must make efforts to develop technology to effect further energy saving, develop technology to better utilize renewable energy and further enhance the safety of nuclear power generation. Furthermore, they must make an appropriate revision to their socio-economic systems.

Is there really a trade-off between environment and economic growth? What principles should guide the behavior of Japan, whose economy is reaching maturity? How should Japanese producers and consumers behave in a mature society?

5 A Prescription for the Resolution of the Trilemma

(1) Revision of the socio-economic system; a transition to a maximum energy saving society through technological development.

(2) Conversion of people's life-styles to those which stress the quality of economic 'metabolism' rather than the quantity of 'throughput'.

(3) Assistance to developing countries in the forms of technical transfers, provision of funds and population control measures.

(4) Establishment of an International Environmental Fund with a provision for special drawing rights; institutionalizing the method of transfer of drawing rights.

(5) Pressing home the point that CO_2 emissions are an act of pollution.

Figure 1
Energy demand and GNP trend

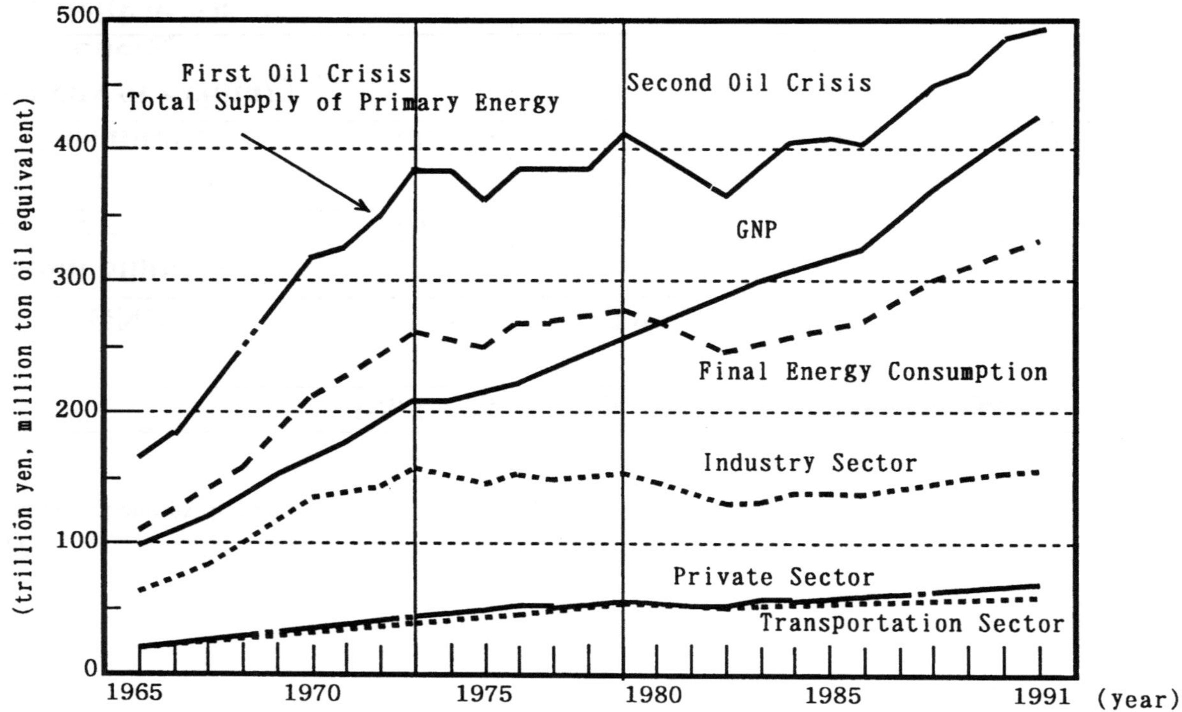

Figure 2

GNP elasticity of energy demand

	1965 - 73	73 - 79	79 - 86	86 - 91
GNP	9.2%	3.9%	3.6%	4.8%
Total Supply of Primary Energy	10.9%	1.1%	-0.3%	4.1%
Final Energy Consumption	11.8%	0.9%	-0.4%	4.1%
Elasticity of Primary Energy	1.18	0.28	-0.08	0.85
Elasticity of Final Energy	1.28	0.23	-0.11	0.85

V.4 Remarks

Dr Levine

No doubt technology will play an important role and we hope it will be helpful, but we are not sure. Regarding Professor Krupp's call for technology assessment, what we need is a technology impact assessment. But it is often misunderstood since we tend to be looking at the impacts of technology and not just at the technology itself. Indeed, technology is one of the crucial elements in attempting to transform society, whatever the problems transformation may bring about. But we are uncertain whether technology will be used primarily for our benefit or not. For technology itself has a dialectic inherent in it, so that once developed it tends to be used beyond the intention.

I think the issue of the elasticity sounds very technical but it is actually very basic. This elasticity of general growth is very important. If GNP is growing at four percent and productivity reflected by energy efficiency is increasing at two percent, then at least we have solved half the problem, i.e. if the energy elasticity is 0.5 percent. At least we keep up the other half by doing other things of great ingenuity.

Dr Sawa pointed out some things that are really crucial and once again they seemed to be lost in the technical details. He noted that the energy elasticity in Japan went down to essentially zero after 1986, but jumped to .85 from 1986 to 1991. This happened not only in Japan, but also in the United States, in West Germany, in Britain, in the Netherlands and in virtually every major industrialized country. The remarkable thing was that it was close to zero in the industrialized world from 1973 until 1986. In this, CO_2 targets are critical to getting back to where we were before 1986, at least as far as the energy system is concerned. That isn't all of the environmental problems, but it's an important part. I think CO_2 targets will help enormously, although I was discouraged to hear that people don't talk so much about CO_2 targets in Germany.

Another comment on Dr Sawa's interesting talk when he discussed putting solar cells on the rooftop and letting the local utility do that, and then after six years reverting ownership to the home owner. I think that is a great idea and I want to suggest what else needs to be done to actually make that successful. The first thing is to be innovative in terms of ownership. The second thing is we ought to talk about storage in a whole new way. It will not be long before it is possible to shed load as a storage medium and so suddenly transform a system from one that cannot be cost effective for a long time to one that can be cost effective a lot sooner. It won't solve the fundamental problem but will contribute to the solution.

I have been asked to give examples of what are energy efficiency technologies. There are a huge number of examples. The first is the conservation supply curve. This shows the cost in dollars per kilowatt, hours per year. When we put the whole U.S. energy system on a chart like this, we end up with a number that is around 40 percent savings that is cheaper than new supply.

What are some energy efficiency technologies that are applicable in developing countries? The Chinese have only worried about industrialization so far, but efficient domestic coal burning saved a million tons of coal equivalent a year. That's a lot, and this was accomplished over a five year period, and it cost less than one quarter of the cost of new supply.

The second is renovation of industrial furnaces. Many of these furnaces had holes, and gas and heat were leaking out of them and they simply patched them up. I think they patched up 60,000 large furnaces over that five year period.

A third one is the use of solid coal waste. What was being done in the early 1980s was that after the coal was collected, there was waste just left in piles. The new technology arose as a combination of an environmental measure to clean up those piles before the crud got into the water stream and also to use what they could of the waste piles for energy.

Processing scrap is number four, where you save a factor of five in iron/steel making if you use scrap, and that was not being done. These two involved improving the quality of coal by washing it. This resulted in a renovation of industrial boilers. Another important method was co-generation. This is very cost effective. Those are some examples and also the general principle.

Professor Huang

Regarding the greenhouse effect in the atmosphere, I think the global and regional cooperation research program and international symposium may play an important role in investigating climate and environment changes. On the energy efficiency, I think industrialized countries should transfer the advanced environment technology to the developing countries.

V.5 Session Discussions

Chairman
Now the floor is open for comments or discussions.

Dr Kashiwagi
An important role of technology and its development is to provide solutions to the trade-off between comfortable life and the protection of the global environment. Yet, most important is to estimate the ratio between electric energy and total energy consumption in the world. This is because the place electricity occupies is increasing rapidly, hence the consumption of those resources for its generation including coal, oil and so on.

When we think about energy conservation technology, we have two categories. One is management of demand and the other is supply-side management. Supply-side technologies are much more important than the demand-side management in my opinion. One example is the solar battery. But it is difficult to get much more market for this technology using only the efforts of the private sector. The public sector should make a greater contribution in this.

Professor Sudara
I try to tell people back home that when they do not have water, it doesn't mean that building more dams will correct the problems. Once you deforest all the trees around there will be no rain and you cannot have water anymore. I do not want us to think that technology alone will help solve the problems. We might get into more and more problems later on. But actually how we use the resources, I think, is more important. We all have to share the world resources on an equal basis. Right now however, the resources have not been used equally. Thailand is a very good case. We have the worst traffic jams possible in Bangkok, but we are importing more cars and building more assembly plants. This is wrong.

Dr Salim (Interpreter)
I would like to ask Mr Sawa three questions. First, on the international environment fund. What's the merit of that compared to the resource purchasing? In the developing countries, we have about 1.2 billion who cannot meet their basic needs. To get the energy for their everyday life is a problem for them. In

fact, almost half the world lacks the energy to carry on their basic life. There are more than 10 million environmental refugees right now.

A critical issue is that where there is gross poverty, introducing the wrong sorts of technology means inducing environmental deterioration. Certainly, poverty is closely linked with the environment. We need to improve the level of poverty as well as addressing the environmental issues. So how should we think about technology relating to the poverty? Looking at the tropical forest in Africa in 1983–1985, there was a great famine in which one million people died. In addition, the past development in the African tropical forests was excessive and did not take into consideration their topsoil, which is quite fragile.

The technologies developed in industrial countries, especially large-scale technologies, may not be adequate or appropriate for the African countries. So when we think about future technologies we need to look for appropriate technology locally. In the past, such technologies were ignored against the large-scale technologies of the developed countries.

In Japan we hear about environmentally friendly technology. Dr Sawa did not touch on this in detail. I would like him to respond.

Dr Nakamura
You mention regulations as being the way to go, rather than deregulation. What does that mean?

Dr Salim
Question to Professor Krupp. Where is the role of culture and ecology? Second question to Professor Sawa. There is a notion of change in consumption. Do I understand correctly that the trend of development will change? You assume that there will be a different type of development in the future and therefore a different technology. In your presentation, you stressed the type of problems developed countries have with the optimizing of consumption and so on. While the developing countries are still struggling with the problems of fresh water, basic foods and so on, do we reconcile the trend of technology in a new type of development with the different demands and needs of the developing countries?

Dr Krupp
Re-regulation is biting its own tail and self-reflecting, so therefore it will be a function of the new transformed system at the same time that it shapes the new transformed system. Basically re-regulation would mean that this, say, selfish individual oriented and materialistic lifestyle is reshaped so that, for instance, cultural values are put to the front and their support for them is for real and not simply mask culture.

Dr Salim, with regard to culture. Now, should the dynamics tell you where culture is going at the moment and will be going in the future? I'm sorry, but I make a difference between Hollywood mask culture and elite culture. To an extent you do support elite culture starting on the indigenous base, in Japan, for instance, which then develops into modern culture. To this extent you can maintain culture which in turn may fertilize the dynamics from the inside. Japan is the most interesting example to see the intoxicating and infectious effects of Hollywood. You can still have beautiful displays, however. For instance, at one concert there was a laser projection against the background so that you saw stars and the Milky Way. But you must have a self-controlled system which can't be controlled from the outside.

Now I come to your question of evolution versus self-reflective auto-control. That is the dilemma of the alternative. On principle, it is possible that the dynamics become self-reflective to the extent that people do know how it works.

Professor Sudara

I don't feel very happy because you are saying that the poverty is a word which everyone is using and they forget it. No, it is not. We are feeling it. But the system is working towards the elite. It is working towards the investor. We would like to be left to use our resources accordingly. But this system was superimposed on us by an elite. This really hurt the people.

Professor Sawa (Interpreter)

Well, I would like to speak to every point. First of all, what I proposed is a global environmental fund with special drawing rights. How is it different from the emission trading? Well, in what kind of merit we can gain from it. The GEF has a higher feasibility than the other because if we can get a consensus among developed countries, then this system will be functional. As to the emission trading issue, what kind of criteria can we establish for the distribution of the emission pyramid among the countries? We can make the distribution proportionate to the population number. Maybe this is the only possible method of distribution. However, it will be very difficult to get to a consensus to this idea. Whether the emission right is void after one year or will be available and valid forever are also points for discussion. In that sense the emission trading has some problems in terms of feasibility. But there is a very important advantage. For example, China will get the emission pyramid 10 times more than Japan. China will start to export emission right. Then, I don't know what the Chinese people will think about it, but the Chinese government or Chinese people may decide to stop the motorization in the 21st century and they may decide to use bicycles only. Maybe they will maintain their own emission. Or they may export the pyramids and then they will get richer.

I say this because this is related to the question raised by Dr Salim. The path of economic development is not a single way. There are multiple or various different paths for development for both developing countries and developed countries and we all should understand that. I think development without motorization is a viable option.

Now I will turn to the next question. What is the opinion of the Kansai Electric Power Company toward my proposition? The first answer I may get is, 'Is it all right for us to raise the electricity price?' because photovoltaic cells are very expensive. The consumer must accept that. But after some time, or once this economy of scale is achieved and as the photovoltaic cells get cheaper, then the burden of the electricity companies will be reduced. Therefore, the price increment will be smaller too. Another point I would like to mention is that photovoltaic cells are the most efficient method of energy production, particularly during the summertime, which is the peak load of electricity demand. For the purpose of peak saving this is a very effective method to implement.

There was a questionnaire survey carried out and the people were allowed to circle what they considered the three main important issues as we move towards the 21st century. Among the Japanese respondents, only three percent of the Japanese respondents circled the item poverty. But, if we carry out the same or a similar survey in the U.S. or France, about 30 percent of the people put a circle around poverty. So I think here it shows the Japanese tend to think locally.

Chapter Six

Scenario Proposal

VI.1 Closing Discussion

Recommendations

The world environment is still worsening steadily. In spite of increasing concern over global environmental problems their improvements do not seem in sight. The world population is growing at the rate of 1.5 percent per year. Energy and natural resource consumption have been increasing with the enlargement of the disparity between South and North. Since the 1992 Rio Summit, little has been done. Despite the pledges given to the protection of global environment, public attention has been distracted toward economic recession on one hand and the ethnic and religious conflicts in various parts of the world.

The realization of sustainable developments should be the top priority of the policies of developing as with developed countries, and, for that purpose, the bottlenecks should be identified, technologies explored and the proper measures, short-term as well as long-term, taken. Against this backdrop, here we raise the following four as the points of recommendation.

1. Technology. There are many promising technologies which have the potential to result in greater energy efficiency. Those utilizing new sources, for example the photovoltaic method, deserve significant support as do others that reduce environment impacts upon lifestyle. Continuous attention should also be paid to the possible impact of CO_2 disposal in one form or another. However, it is of great importance to carry out complete technology impact assessments prior to the application of new larger scale technologies.
2. Policy Instruments. In order to stop further degradation of the environment there is an urgent need for a shift in the environmental policy approaches to enable these to be more effectively implemented. The emphasis has to be shifted from the traditional command and control approach to those economic policies which transform the market. By giving price signals through the imposition of taxes or charges on environmental damages, policy makers could encourage polluters to change their behavior and to put theory into practice. Moreover, environmental taxes could make pollution avoidance and its minimization profitable, produce significant savings in pollution abatement costs, and support further innovations. By replacing the taxes on the use of labor with environment taxes one could move the economy onto an environmentally sustainable and job creating path.

Apart from shifting to environment taxes, particularly on the use of energy, governments should also encourage the application of economic incentives such as environmental labelling, environmental auditing, the public's right to know, user charges and tradeable permits. Environmental quality standards have to be paid more attention. The efficient emission reductions, however partial they might be, do contribute to the efficiency gains. This is often more than offsetting an increase or shift in total demand. Therefore, as in the CFC and CO_2 cases, total emission ceilings are indispersible, since the traditional approach of direct emission control has reduced cost-effectiveness. There is also the urgent need of the internalization of environmental criteria into policy areas such as transportation, energy, industry, agriculture and tourism from the very beginning. By such preventive measures, economy is going to be more sustainable against the ever increasing burden of pollution and clean-up costs.

3. International Collaboration. International collaboration between South and North, beyond the east and west parallels, should be strengthened. Particularly, aid should be restructured so as to assist the developing countries in implementing more energy efficiency policies, including training programs. More emphasis should be laid on technological cooperation between the developing and the developed countries and should allow the former to improve their own technology. Technology transfer has resulted in commercialization of high technology which tended to serve the interest of the developed countries. International trade is no longer dominated by the developed countries alone. The pricing of traded goods which are distortedly weighted in favor of the developed economies should be restructured. The common resources, including land, sea and air, should be of equal concern. Developed countries should not exploit such resources without serious consideration of the impact on developing countries.
4. Steps toward Sustainable Development. In the aftermath of the Cold War, with the restructuring of the global economy under way, industrial nations should expand their role, working with the developing nations in seeking a sustainable development path. The present course to reduce economic cooperation in light of continued economic recession is of course short-sighted. There is a strong need to reverse the course. What is really needed are initiatives by industrial nations to support the wider application of energy efficient technologies in developing nations. It could have a huge impact globally. Such a program needs to create expertise in energy efficiency through training, institution building, and supportive policy making that demonstrates the feasibility and accountability of large-scale efficiency projects. This, with a sufficient investment, would help developing countries lead the whole process of industrialization onto the energy efficient path. Industrial nations should direct one to two billion U.S. dollars per year into aid for energy and 20 to 40 billion dollars per year of capital investment into those industries that implement energy efficiency projects. With this alone, the overall energy growth in the developing world could be cut in half with larger environmental and economic benefits to all parties.

VI.2 Postscript: Scenario for the Future

Aftermath of the Symposium

As seen in the previous chapters, the symposium dealt with an extremely wide range of issues concerned with the global environment. This is partly because we, the organizing committee, intended from the onset to organize a symposium in which the 'interface' of different expertise could take place. In this regard, as intended, the symposium was quite a success. This is also partly because we wanted cross-discussions, if you will, beyond the divisions of natural and social sciences, of theory and practice, of South and North, and so on. Such an attempt was difficult, if not impossible, due to the depth of the divisions in the above. In this too, the symposium was a remarkable success. But most important was the fact that the sponsor of the symposium, IBM Japan, was willing to offer such an opportunity to those concerned with the global environment. This makes the symposium quite unique.

To synthesize those discussions, some of them conflicting, into a book seemed, in the aftermath of the symposium, to be an almost impossible mission, however. In the beginning of the editing session, following the symposium, we wanted to present the papers and discussions as they were. Then, during the course of editing works for publication, we decided to reorganize the whole discussions in a more theoretical manner. This was because the symposium brought about a number of important 'discoveries' that require further research. Such an attempt however proved to take considerable time and effort. As a consequence of the ever pressing time factor, we have determined to present the symposium as it was.

Indeed, the speed with which technologies are developing is rapid and the range and scope of environmental studies is fast expanding. Above all, many more governments and non-governmental organizations have begun to put their policies and programs into action. Yet the contents of the symposium remain extremely valuable and valid today.

From the Holistic Analyses to the Concrete Actions

When we came to recognize that many of the nations in the world were still struggling for growth as the only answer to staggering poverty, it was still a long way to sharing the messages of the Rome Club. Only a decade after, the serious search for alternative development began to take root. The Brundtlant Commission put forward sustainable development as the concept which every nation can share. However, another two decades were spent in making the concept acceptable as the guiding policy toward both development and the protection of the global environment. By that time, more than 100 nations, South and North, had agreed to implement domestically such concrete action programs as Agenda 21, adopted at the Rio Summit.

This history reveals the victory of a holistic approach to the issue of environment. It enables us to look at each environmental problem in a wider perspective. It also leads us to a better understanding of environmental problems from varied scientific viewpoints. Thanks to such an approach, it has become much clearer that the protection of the global environment requires global collaboration. Yet, as the discussions we had two years ago revealed, we still need to scrutinize technological breakthroughs such as CO_2 fixation, photovoltaic cells and so on. Information on such developments are always in demand and it is important to encourage the so called technology transfer from North to South.

Yet it would be safe to say that since the Rio Summit, with or without such technological breakthroughs and collaborations, many more governments and citizens are organizing toward finding solutions to the existing problems. Of course, the problems and the solutions they choose depend upon the country, region or even community to which they belong. In other words, environmental problems can no longer be confined within academic arguments. They should be seen as part of daily life to be solved on a day-to-day basis. Two cases in this volume amply indicate how much a small-scale effort can achieve, namely the efforts of the city government and citizens of Kawasaki to delimit toxic emissions, and of the prefectural government and people of Shiga to clean up Lake Biwa. This suggests that the time has come to combine the accumulated knowledge of the environmental sciences and technologies with the specific needs of nations, regions and communities so as to stop the continuing degradation of eco-systems at the grassroots.

Toward Environmental Security

The change of perspective in looking at the global environment also reflects the recent political changes, namely the end of the Cold War. It marked the end of ideological disputes and at the same time the beginning of transformation of the existing world systems. Of particular importance here is the conceptual change of environment. Lester Brown talked about this, saying that 'building an environmentally sustainable future depends on restructuring the global economy, making major shifts in human reproductive behavior, and bringing about dramatic changes in values and lifestyles'.

From this, as many arguments in the discussions of the symposium show, it follows that conceptual transformation of our economic, social and political life is as necessary for a sustainable future as is technological development. However, the transformation cannot take place overnight. Then what should be done?

If we look at the transformation affecting the concept of security we see that the emphasis is changing from being focused on the protection of nations toward the creation of common peace. Following such a line of transformation, it would one day be necessary for those concerned with the protection of environment to move toward a concept of environment based on human security. We do see a new horizon for future researches on preventive measures. Cutting across national boundaries, we now need to inquire into the possibility of environmental security. This is, I believe, one of the most significant results of the symposium. In other words, as the concept of security comes increasingly nearer that of environmental protection, we need to develop a concept of environmental security.

FSC
www.fsc.org
MIX
Papier aus verantwortungsvollen Quellen
Paper from responsible sources
FSC® C105338

If you have any concerns about our products,
you can contact us on
ProductSafety@springernature.com

In case Publisher is established outside the EU,
the EU authorized representative is:
Springer Nature Customer Service Center GmbH
Europaplatz 3, 69115 Heidelberg, Germany

Printed by Libri Plureos GmbH
in Hamburg, Germany